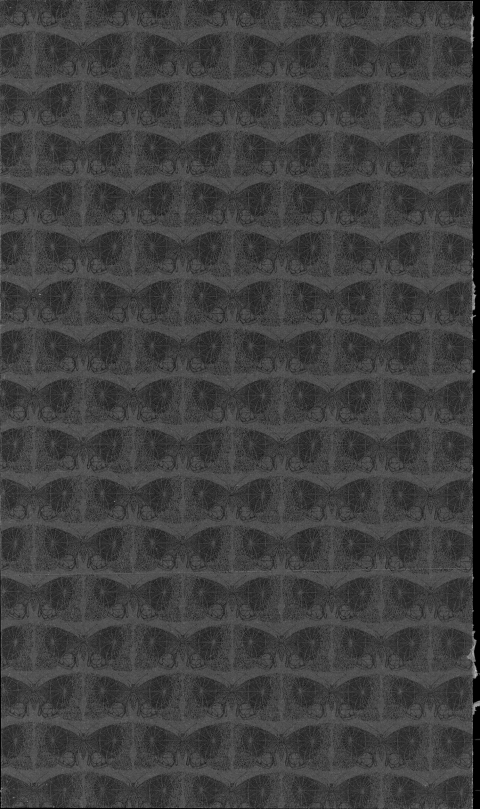

A Little History of Mathematics

INSPIRING GUIDES FOR CURIOUS MINDS

Whether you know absolutely nothing about a subject or are already familiar with it, these *Little Histories* are the most energetic, entertaining and reliable guides you will find.

A Little History of the World by E.H. Gombrich
A Little Book of Language by David Crystal
A Little History of Philosophy by Nigel Warburton
A Little History of Science by William Bynum
A Little History of Literature by John Sutherland
A Little History of the United States by James West Davidson
A Little History of Religion by Richard Holloway
A Little History of Economics by Niall Kishtainy
A Little History of Archaeology by Brian Fagan
A Little History of Poetry by John Carey
A Little History of Art by Charlotte Mullins
A Little History of Music by Robert Philip
A Little History of Psychology by Nicky Hayes
A Little History of Mathematics by Snezana Lawrence

New titles coming soon!

Discover more about the full series at:
yalebooks.co.uk/little-histories   yalebooks.com/little-histories

A Little
History of
Mathematics

Snezana
Lawrence

YALE UNIVERSITY PRESS
NEW HAVEN AND LONDON

For information about this and other Yale University Press publications, please contact:
U.S. Office: sales.press@yale.edu yalebooks.com
Europe Office: sales@yaleup.co.uk yalebooks.co.uk

Set in Minion Pro by IDSUK (DataConnection) Ltd
Printed in Great Britain by Bell & Bain, Thornliebank, Glasgow

Library of Congress Control Number: 2025930627
A catalogue record for this book is available from the British Library.
Authorized Representative in the EU: Easy Access System Europe,
Mustamäe tee 50, 10621 Tallinn, Estonia, gpsr.requests@easproject.com

ISBN 978-0-300-27373-1

10 9 8 7 6 5 4 3 2 1

Contents

CHAPTER 1

Mathematical Imprints

How does mathematics make you feel? For many people, mathematics can seem forbidding: too hard, too cold, too abstract. That sense of dread may have started at school, especially when mathematics lessons were a matter of manipulating symbols and doing obscure calculations. Or it can be inherited, or passed on by others – if people around you are talking about maths being diffi-cult or pointless, you're very likely to think the same. And so, lots of people drop mathematics as soon as they're able. 'What's the point of it?' they ask. It doesn't seem relevant to anything you might need in the real world.

Others, though, experience something like an intuitive under-standing of mathematics which they nurture over time. Maths is difficult, yes, but these people embrace the challenge. They can see a beauty in mathematics that others – sadly – can't. Maybe they are interested in having a career which requires mathematical knowl-edge. They develop a love of the subject and are excited by the places it can lead them.

Whether you are fearful of or fascinated by it, the history of mathematics will show you not only a different face of mathematics, but exactly how and when it has been useful. How beautiful it is sometimes, too. As we shall see, mathematics is not just about numbers. It is about being able to work out some kind of general rule and then how it can be applied to other situations. Mathematics develops skills that can solve problems of all kinds, mathematical or not. In this little book you will find, through learning about its history, many such examples of the versatility of mathematics, far beyond equations and arithmetic – from the surface of the Earth all the way to the stars.

Perhaps at some point you have seen some mathematical writing and not understood it. You would not be the first; rest assured, even professional mathematicians sometimes have to rely on discussions with colleagues to properly understand problems they are looking at. But how do you recognise some writing is mathematical in the first place? The complicated notation that might spring to mind – all those strange dashes, squiggles and letters – are obvious signs but, as we shall see, a lot of that is really quite modern. Mathematics has been going on for a long time before the dashes and squiggles were invented. Put simply, there has to be something *mathematical* going on for us to say that it is mathematics. And if we are dealing with writing from a very distant past, in a language that is not familiar to us, from a time even before recorded language, that can be sometimes difficult to recognise.

Such was the problem faced by Belgian archaeologist Jean de Heinzelin de Braucourt when in the 1930s he stumbled upon some ancient treasures near Lake Edward at Ishango (now in the Democratic Republic of the Congo). Among the human remains and stone tools, he discovered a sharp-tipped baboon bone, about the size of a pencil, with strange etchings on its sides. Later excavations at the site uncovered more evidence – weapons, spears, and even some old ropes – that proved this was a place where a prehistoric group of humans lived. The star of the show, though, remained the Ishango bone. But what was it?

The 168 etchings on its sides seemed mysterious, but suggested something more than mere decoration. Having to work things out from marks or symbols is often a good sign that what we are dealing with is mathematical. And so it proved. The Ishango bone turned out to be some kind of recording or counting device. The inscriptions on it were a form of mathematical statement. The etchings were organised into three parallel columns, each showing a different procedure: the doubling of values; listing prime number values between 10 and 20; and summing two values around the number 20. Prime numbers are those you can only divide by 1 and themselves. As we shall see, they were first described around 500 BCE, many centuries after the Ishango bone was made around 20,000 years ago. Of course, some may say these marks were a happy accident, but they are just a little too organised to be the result of random scribbling. We now largely agree that the Ishango bone is the earliest mathematical object in the history of mathematics to have been discovered – so far.

To call an object mathematical, it has to be made or shaped by a human effort, or contain or refer to something mathematical. In an everyday sense, we would consider the Ishango bone, or an abacus, or a calculator, all to be mathematical objects. But a mathematical object can even be an idea of an object. Not all mathematical objects, in other words, are those you can touch, like this bone. Mathematicians nowadays call mathematical objects all the items we can either act on or do something to. For example, a point A, a line a, a number n, are all such objects. These are abstract objects: when mathematicians refer to a point A, for example, they refer to a mathematical point, with certain characteristics (perhaps its position), and not usually to a real point in a real space.

We may therefore have stumbled upon our first difficulty. Or perhaps we are getting closer to understanding what a mathematical object is. The Ishango bone is not an abstract object – it is a very real thing – but what it contains is quite abstract, and it refers to some knowledge we can certainly call mathematical. Its unique inscriptions show the first counting system of a kind, recorded by humans. That is what makes it mathematical.

Does mathematics exist independently of us? Mathematical laws and rules are somehow inscribed in nature and the universe. Some animals appear to behave mathematically: for instance, magpies can tell the difference between five and six objects; chimpanzees can give a nod to the fact that five is more than four. And let's not forget the bees, crafting perfect hexagons, or ants, who are quite capable of spotting exits from enclosed spaces. But this seeming mathematical ability is no more than recognition, instinct and inbuilt behaviour. They don't, in other words, build on that behaviour further. Bees don't connect their honeycomb structure with ideas about the best way to pack things in general (that was done, as we'll see, by a twentieth-century mathematician). They just do it.

We humans always try to make things better, and when we manage to do something well, we still seek to improve it. Sometimes that leads us to places where what we imagine and structure with mathematics is not possible to reconcile with our reality. This is very different to what bees and, as far as we know, any other animals do. For over two millennia, humans have built on their knowledge of recognising patterns and connecting them with other patterns. Now we can use mathematics to postulate many more dimensions than the three we live in. With mathematics we are able to make predictions in seemingly chaotic behaviours that occur in all sorts of fields, from nature to economics. We have developed abstract structures of mathematics so high that if you stood on them and looked back at the Ishango bone you would feel quite dizzy. And we continually push that knowledge on.

We have just touched on some of the things that make up the two branches of mathematics: pure and applied. In pure mathematics we make abstract conclusions. This sort of maths includes number theory, geometry, algebra, analysis, all of which we'll find out more about later in this book. What we learn from pure mathematics we may not use immediately, or ever, directly. Sometimes we wait for centuries before we can see where we can use such knowledge in a practical way.

Applied mathematics, on the other hand, comes usually from some problem we want to solve in real life. And we would usually use such mathematics, in real life, more or less immediately. Applied mathematics is involved with questions and problems that arise from other disciplines, such as physics or engineering, computing, biology, economics – you name it – and it also contributes to the developments in these disciplines.

One thing pure and applied mathematics have in common is the formulating of rules and the way of making generalisations. If applied mathematics can help make sense of the patterns of majestic whales and their mesmerising aquatic dances across the vast oceans, pure mathematicians are like deep-sea divers immersing themselves in the fascinating world of ideas just for the joy of it. But even when this world doesn't seem to have anything to do with our everyday experience, they are keen to describe its laws and structures. And through the history of mathematics, be it pure or applied, we can follow such adventures from a distance yet still share in awe and wonder of their discovery.

Humans in different times and places have always glimpsed new things and come up with new ideas, recorded them, and played with them, and so created new mathematics. It's like language: just as we can bend words to our will in creating astonishing new poetry, while retaining enough linguistic structure to be comprehensible, so too we continuously update and originate new mathematics out of that which we inherited. This is another reason that the history of mathematics is important: looking at successive breakthroughs in the field, we can learn of the crucial axes upon which mathematics as a whole has been founded, and the turns it has taken on its journey of development. We can see the dynamic dance of mathematical ideas between peoples far away from each other, not only in space but in time too. Through the mathematics they write, universal dialogues develop and problems that were posed in one century may be solved in another. Mathematicians are always answering questions of today by building on what has been discovered yesterday.

So, from the imprints made on the Ishango bone, get ready for a journey on which we will explore many more fascinating mathematical 'imprints' left by various cultures over many centuries.

Throughout this book, we will marvel at seeing how mathematics evolved, how humans honed different skills in different cultures to inject new ways of thinking into the discipline. In this historical adventure we will not only decipher such mathematical imprints but will also understand where they originally came from, and the vital contributions that men and women all over the world have made to mathematics.

Suddenly, even the most mundane things you do in your everyday life may become infused with some mathematical magic. When you go to the supermarket, some historical mathematician may help you figure out the best deals. When you're budgeting your cash, it will be medieval mathematics that will help you make the best decisions. Doing your taxes, once you get down to it, may be less of a headache with a bit of maths know-how. I can't promise it, of course, but all of this might change your outlook not only on maths but also on the world. You may become a pattern-spotting, conclusion-drawing whizz. You may feel empowered in understanding mathematics you never thought you'd be able to. You may even feel you have a superpower of a kind, as the mathematics underlying so much of our modern world is revealed to you. One thing is certain, you will be able to appreciate the history of mathematical discoveries and hopefully enjoy more doing mathematical procedures in your everyday life.

It is often said that 'mathematics is the language of the universe'. Let's spend some time together exploring this language and its many ways of describing the world we live in. Through its history we can also put ourselves in the shoes of past mathematicians and see how they came to formulate and expand the knowledge they inherited. In doing so, they changed their reality ever so slightly, and brought it closer to our own.

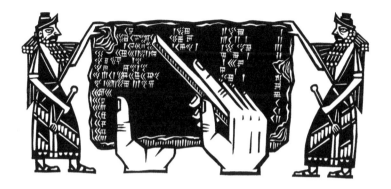

CHAPTER 2

Unearthing Wisdom

In 1922, George Arthur Plimpton, an American publisher and collector of old books and antiquities, bought a little clay tablet for $10. It contained some mysterious writing that at the time only a handful of people in the world were able to read. Who would have thought that this tiny tablet would turn out to be one of the most celebrated mathematical treasures of all time?

Plimpton bought this tablet, now named Plimpton 322 after him, from archaeologist and adventurer Edgar J. Banks. Something of an Indiana Jones, Banks was the American consul in Baghdad, and used to buy many such little treasures there and in the markets of Constantinople to sell to wealthy collectors back home. He might have thought he got a fair price for the postcard-sized tablet, but it is priceless today. For much of what we know about ancient Mesopotamian mathematics comes from deciphering this little object.

Measuring 13cm high, 9cm wide and 2cm thick, the tablet was crafted around 1800 BCE, and was likely found somewhere in what is now southern Iraq. It is packed with writing in a sharp-edged script. The inscriptions are arranged in four columns. If you read it like you would an English book, from left to right, you would notice that the first column has a list of numbers listed in descending order. These numbers have a special relationship with two other numbers in the same row. But, if you read it from right to left, like one would Hebrew or Arabic, the first column turns into a list of ordinal numbers (1st, 2nd, 3rd, and so on) that label each row. How do we know these are numbers? Well, people didn't know that for a long time, as the languages of the Mesopotamian era were lost for many centuries. Thankfully, by the time the tablet was brought to the United States, these languages had been identified and could be deciphered.

The Mesopotamian civilisations – including the Sumerian, Assyrian, Akkadian and Babylonian – were geographically defined by two large rivers, the Euphrates and the Tigris, and flourished in the period 3100–540 BCE. This 3,000-year span is an astonishing endurance – longer than all the time that has passed since their demise. It is somewhat strange, considering how long these civilisations endured, and how well known other ancient civilisations are to us, how little we hear about the ancient Mesopotamians. It is also quite amazing that until the middle of the nineteenth century, their achievements lay mainly unknown because the inscriptions that described them were indecipherable.

The cuneiform script, invented in Mesopotamia, is in fact the earliest script we have evidence for in the history of the world. Like Latin later, cuneiform was developed and used in many different cultures with different languages across this vast region in the period. As scholars started to decipher Mesopotamian texts, they realised that the predominant number base used to count and calculate in was 60, called *sexagesimal*. We in fact still use this when we measure time. It was in ancient Mesopotamia that the day was divided into 24 hours, each hour into 60 minutes and each minute into 60 seconds. This system survived for over 4,000 years. But we also use this system in mathematics when we measure angles. There are 360 degrees in a circle, 60 minutes of arc in a degree and 60 arcseconds in a minute.

Let's examine our little Plimpton 322 tablet in more detail. Numbers are written all over it, in four columns. These should be read from right to left, beginning with the number of the row. The remaining three numbers in each row are linked to each other. They represent three values related to right-angled triangles. How can that be, you may ask yourself? We know, and the writer of Plimpton 322 also knew, that in a right-angled triangle, the square you can draw on its longest side (called the *hypotenuse*) will be equal to the sum of the squares on the other two sides. This is now called *Pythagoras' theorem*, named after the Greek mathematician we will meet in Chapter 4.

You can therefore use any aspects of such a triangle – the length of its sides, or the area of the squares constructed on its sides, or any ratio between these lengths and areas – to list three values that refer to a triangle that is right-angled. We will shortly see how a choice of such numbers was organised in each row. But first we need to look at some mathematics to understand why this tablet is so unique and important, apart from giving us measurements with which we could construct right-angled triangles.

We will start with the prime numbers, numbers that are divisible only by themselves and 1. Every whole number can be expressed as a product (the result of multiplication) of its prime factors. A factor of some larger number is a number that can divide the larger number without remainder. To see how that works in practice, let us take the number 72. You can break it down into its prime factors and write it as a product of its prime factors, $2 \times 2 \times 2 \times 3 \times 3 = 72$, or $2^3 \times 3^2 = 72$. Primes are sometimes called the building blocks of the universe of numbers. This means that we can deconstruct all whole numbers back to their building blocks through finding their prime factors.

There are certain numbers whose prime factors are *only* 2, 3 and 5. Such numbers are called *regular sexagesimal*. A regular sexagesimal number will be a divisor of any number which is a power of 60 (like, for example, $60 \times 60 = 3600$). A divisor of a number can divide that number without remainder. In other words, a divisor is also a factor of a number. This also means that you could take 60 to any power, and you will know it would be divisible by any number that one can make by multiplying 2, 3 and 5 (and powers of these) to make it. The inscriber of the Plimpton 322 tablet didn't say they knew all of this, but they certainly used the knowledge of it. That is because all the numbers from the two middle columns on the tablet are regular sexagesimal numbers. In other words, you can construct them by multiplying the powers of prime numbers 2, 3 and 5 *only*. The two middle numbers in all the tablet's rows are also related: one represents the length of a shorter side of a right-angled triangle, and the other its hypotenuse. The last column (from right, or first from left) is a ratio of the squares of these two sides.

That is quite amazing, considering that we don't have much evidence of mathematics before this tablet, not to mention that the proof of what we now call Pythagoras' theorem came only many centuries later.

Although it is probably the most famous, Plimpton 322 isn't the only tablet that has come down to us giving an insight into Mesopotamian mathematics. Others from this period, called the Old Babylonian, covering roughly two hundred years between 1800 and 1600 BCE, also show tables of results that could be used in calculations when they were needed – perhaps to construct right-angled triangles, or to know the ratios of the areas of squares on such triangles.

Mesopotamian mathematicians were very good at, and very keen on, making tables displaying lists of numbers with the same repeated operation or calculation made upon them, and the consequent results. We find other clay tablets with multiplication tables, division tables and *reciprocal* tables. A reciprocal is a number that, when multiplied with the original number, gives 1 (for example, 2 multiplied by $\frac{1}{2}$ equals 1 – so $\frac{1}{2}$ is the reciprocal of 2), and there are Mesopotamian tablets displaying numbers and their reciprocals alongside each other, to use in calculations more readily. The Mesopotamians also knew how to obtain a square root of a number (that is, the number that when multiplied by itself gives the original number). Today, this skill is almost forgotten as we mainly use calculators to work it out, but doing it by hand involves a series of approximations generating results closer and closer to the required result. As this calculation involves quite a lot of repetition (called *iterations*), it could take a long time to complete. It is very useful therefore to have a table of precalculated square roots of certain numbers. The formula of how to get square roots of numbers is sometimes now attributed to a much later Greek mathematician, Heron of Alexandria (around first or second century BCE). But there are no explicit formulae given in these Old Babylonian tables, and neither do we have the repetitive results needed to deduce the way that the results were obtained. This is another major feature of Mesopotamian mathematics more generally: we know that they knew how to do things, but we don't know how they knew it. They did not leave any proofs to show how things work or why they do.

The majority of the clay tablets from this period – hordes of which have been unearthed in present-day Iraq – are believed to have originated from schools, or certainly from some kind of group of learners. Many of the mathematical problems are posed in words, rather than as we expect them now, in symbols, and are almost always to do with the computation of a number. Unlike mathematics from later periods, when the emphasis was on students being asked to prove something, the clay tablet problems pose specific questions relating to a measurement of some kind. So, for example, we find questions about how to calculate the length of a canal, areas of fields, or the number of bricks used in constructions. The presence of answers tells us that some kind of learning was taking place here – which is why we deduce that most of such groups of tablets come from schools of some kind.

As many of these problems are to do with real-life situations, understanding the units of measurement is very important. We know that the Mesopotamian sexagesimal system generally means that different units are simple fractions or multiples of the base 60, and we can translate it into our own metric system – so, for example, the smallest unit of measurement was the *she* (barleycorn length), which is about $\frac{1}{360}$ of a metre. But looking at Mesopotamian mathematics as a whole, working out the units of measurement used at any particular juncture is by no means an easy thing for us to achieve. The measurement systems of ancient Mesopotamia were very complicated and, inevitably, changed over the long span of the civilisations. Just looking at the relatively limited Old Babylonian period – from which the most interesting and sophisticated mathematics comes – we find a very rich diversity of units.

Unfortunately, despite all the clay tablets that have been unearthed so far, we still don't know who made or wrote them. We can't even tell whether any of the authors were mathematicians: scribes, students or even teachers. Some of these tablets certainly represent original findings – the first time that someone wrote down something important, which others then copied – but usually we can't tell which these are with any certainty. We can guess that mathematics was used in Mesopotamian cultures quite often as there are so many tablets like Plimpton 322 – it seems that there were many people who studied mathematics and knew how to apply it in practice. Looking at the literature of ancient Mesopotamia, we can also determine that its peoples were passionate, brave and very funny at times.

Mesopotamians gave us the first written script, the first tables to speed up calculations and the first way of measuring time. They showed each other how to calculate things, but didn't prove why or how things worked. That came with the Greeks, an ancient civilisation we will visit shortly, but not before we make a visit to ancient Egypt.

CHAPTER 3

A Scribe's Story

Discovered in Thebes within some dusty ruins around 1858, one papyrus would become probably the best-known and most valuable of all artefacts telling the history of Egyptian mathematics. An antiquary and traveller, Alexander Henry Rhind, had purchased it a few years before he died, and the papyrus came to be named after him when it was sold to the British Museum after his death.

The Rhind Mathematical Papyrus, along with a leather scroll also purchased by him, a few other papyri (one of which is in Moscow) and some wooden tablets: these are pretty much all the historical objects that have come down to us that give direct evidence of the mathematics developed by the ancient Egyptians during their long and productive three-millennia civilisation. Needless to say, our knowledge of Egyptian mathematics is therefore pretty limited, and perhaps not surprisingly, this era is often rather overlooked by historians of mathematics. But how could the Egyptians construct such magnificent architecture – those astonishing pyramids, for example – if they weren't mathematically skilled? We can only catch glimpses of what must have been a keen mathematician's understanding of engineering, construction and design.

Compounding the difficulty, the world lost the ability to read Egyptian hieroglyphs at all for many centuries. By the time Egypt was conquered by Assyrians (671 BCE), Persians (525 BCE) and finally the Greek Ptolemaic dynasty (332 BCE), there was little left of the original culture. Towards the end, Greek and demotic scripts were used side-by-side with the Egyptian, and by the time Romans replaced the Greeks as a predominant civilisation in the Mediterranean, hieroglyphs were mainly only used by Egyptian priests. It was not until the early nineteenth century, after the Rosetta Stone (now in the British Museum) was found near Alexandria in Egypt, that French philologist Jean-François Champollion deciphered hieroglyphics, and so the reading of Egyptian mathematical texts and their analysis could finally begin.

The scant mathematical sources that have come down to us all date to the Middle Kingdom, from about 2055 to 1650 BCE. The mathematics therein fall into two broad categories: on the one hand, useful collections of data that could be used for calculations, similar to what the Babylonians left us; on the other, texts that show how to solve practical problems, such as dividing property or measuring the area of a field, by using mathematics.

Contrary to the Babylonians, the Egyptians established the use of the base 10, which we still use in everyday mathematics today. Egyptians didn't have a sign for zero – that came later, as we shall see – but noted 'nothing' with an empty space. They also didn't have positional notation like we do. For example, in the number 1,234, each digit has a very precise meaning based on its position. Ancient Egyptians used different symbols to signify units – the tens, hundreds and so on – and repeated them to show how many of each unit a particular number contained. Take 234,567 for example – here given from left to right, though ancient Egyptians mostly carved or painted their hieroglyphs from right to left:

= 2 tadpoles (100,000) + 3 fingers (10,000) + 4 lotus plants (1,000)
+ 5 coils of rope (100) + 6 cattle hobbles (10) + 7 strokes (1)

The mathematics on the Rhind papyrus was written not in hieroglyphics but in the cursive hieratic script. This looks slightly different from hieroglyphics, being less pictorial, and was developed around 3000 BCE to make life a little easier as using it didn't require as much skill. The writing in hieratic was usually done in ink using a reed pen on papyrus, leather or wood. It was used for longer writing and documents, while hieroglyphic script was reserved for carving on stelae, walls and tombs.

The Rhind papyrus, around thirty centimetres high and almost two metres long, contains sections on arithmetic and algebra (where Egyptian fractions, which we will see shortly, appear), geometry (with descriptions of how to calculate areas and volumes of some three-dimensional figures, such as pyramids) and various other interesting little problems. The Rhind Papyrus is incredibly important not only because it gives us a rare insight into Egyptian mathematics, but because here, for the first time in the subject's history, we find someone who did some mathematics – the scribe who wrote it signed the document. His name was Ahmes (sometimes transliterated as Ahmose) and he wrote this papyrus around 1650 BCE. He very clearly states that he isn't the author of the original text, but a scribe, someone who transcribed the original. The text, he said, came from an earlier work of about 2000 BCE, so about 350 years before he sat down to write his roll. Now, if you or I wrote a book based on a seventeenth-century mathematical text, we would be called mathematical historians rather than mathematicians. We can't tell if Ahmes simply directly transcribed the text, or if he improved or changed it. What he does do in the opening of this scroll is to present it as something that will give a reader a 'correct method of reckoning, for grasping the meaning of things and knowing everything that is, obscurities and all secrets'. He certainly knew how to entice a reader.

The sources suggest that the Egyptians favoured calculations using halving and doubling. They also almost worshipped fractions! They liked to look at a fraction and then decompose it into ever smaller fractions. These smaller fractions would be unit fractions, in other words, they had a unit (1) as the numerator (the top of the fraction), which was generally scribed as an oval-shaped 'mouth' hieroglyph, and another number as its denominator (the bottom of the fraction), which would be a group of hieroglyphs representing the number (four strokes for the number 4, for instance). In this way, though they lacked a means of notation for writing fractions like $\frac{3}{5}$ or $\frac{6}{7}$ as we do today, ancient Egyptians could write any fraction as a sum of unit fractions. This could be practically useful. Let's say you're an ancient Egyptian landowner trying to divide a dozen pieces of land between seven workers, so each worker would get $\frac{7}{12}$ of the total area. Being able to express it as a sum of two unit fractions, such as $\frac{1}{2}$ and $\frac{1}{12}$, would, make this much easier: each person gets $\frac{1}{2}$ and then $\frac{1}{12}$ of a piece of land.

There were various rules used to decompose fractions into further, ever smaller fractions in a variety of problems that Egyptians left to us in their writing. This extended to units of measure too: for example, a unit of volume, a *heqat* (about 4.8 litres), was divided into ten parts, called *henu* (each about 0.5 litres). In general, with fractions, Egyptians would progressively halve them from larger to smaller fractions, for example $\frac{1}{2}, \frac{1}{4}, \frac{1}{8}, \frac{1}{16}, \frac{1}{32}$ and $\frac{1}{64}$. One can always go into further smaller subdivisions – the smallest subdivision in the case of volume was *ro*, corresponding to $\frac{1}{320}$ *heqat*.

These unit fractions are believed to have been related to Egyptian mythology, and are sometimes therefore called Horus-eye fractions. You may have heard of the Egyptian Book of the Dead – the text that deals with the work that needs to be done after death to enable the soul of the deceased to transcend this world and travel into the next. In the 112th chapter, it tells the story of Horus, the son of Isis and Osiris, two great gods in the Egyptian pantheon. Sometimes Horus is used also as one of the names of the Sun, and of course, where there is light, there must also be darkness. The god of darkness, disorder and destruction was Seth, an unlikeable character often portrayed as a black boar. Seth attacked Horus, slew him and swallowed his eye. Following this gory event, the god of mathematics appears, named Teth or Tehuti. He was often portrayed as a baboon, a clever monkey, and was the measurer of this and the other world. He was said to 'have calculated the earth and counted the things which are in it'. Teth was also a defender of light, and he revived Horus, restored his eye and (permanently) chained Seth.

Horus' eye is portrayed as an iconic symbol of light's victory over darkness. You can still come across this symbol in amulet form which is said to be a sign of protection. But that's not all. There is a mathematical myth that connects the eye of Horus with Egyptian unit fractions. If one adds the fractions denoted by the hiero-glyphics forming the image of the eye of Horus, the result is $\frac{63}{64}$, missing therefore $\frac{1}{64}$. Does this mean that some things can never be restored fully? Or that seeing and knowing something can never be complete? Befittingly, this is one of the many stories about Egyptian mathematics that we can't be certain about.

We'll see later that Greeks were interested in Egyptian mathematics – which should come as no surprise, given that what was to become Greek mathematics often arose geographically in northern parts of Egypt, with Alexandria as one of its centres of learning. Indeed, the ancient Greeks believed that mathematics was invented in Egypt. Proclus (412–485 CE) and Herodotus (c.485–425 BCE) wrote that because the annual flooding of the Nile brought both life and destruction to the people who lived on its shores, Egyptians needed a system that would help them record and re-establish the boundaries of their lands after the waters had receded. It's an idea that has been repeated many times since. Aristotle (384–322 BCE), meanwhile, believed that geometry was borne through the discus-sions of Egypt's priestly, and therefore leisured, class.

One of the other popular stories about the origination of math-ematics in Egypt concerns the rope-stretchers, or *harpedonaptai* – people who were tasked with measuring land and making constructions using knotted ropes. Images of *harpedonaptai* decorate many monuments from ancient Egypt, so it was an important role in Egyptian society. They would, for example, use a rope with twelve segments to construct a triangle that is right-angled, where sides would be of 3, 4 and 5 segments. We have already seen that Mesopotamians knew how to construct right-angled triangles, as demonstrated on Plimpton 322; Egyptians did this too, in a different way. Such stories about Egyptians were first told by the Greek philosopher Democritus (c.460–c.370 BCE), so about 1,000–1,500 years after images of the rope-stretchers were first inscribed. There was even, apparently, a ritual of stretching the rope, involving an elaborate ceremony to mark the foundations of new buildings. In a way, we still do that when we cut a cord to open an important new building.

Alas, no new ancient Egyptian mathematical artefacts have come to light since the discoveries of the few sources we now rely on and have explored here. But an interesting thing did happen in more recent times: Egyptian fractions became a matter of interest to modern mathematicians – number theorists. Number theory is a branch of mathematics that deals mainly with the properties of positive integers (whole numbers), and we'll see much more on this later in this book. Mathematicians were looking at *rational numbers*, which are numbers that can be written as fractions: familiar things like $\frac{1}{2}$ and $\frac{1}{3}$, but also $\frac{63}{1}$ – it's just we don't usually write '1' in the denominator. It was shown that every rational number has *infinitely* many distinct Egyptian fraction decompositions. In other words, you can break any rational number in infinitely many ways starting from a larger and going to smaller unit fractions. Those long-ago scribes are continuing to speak to us.

CHAPTER 4

The Secrets of the Pythagoreans

Even if you know next to nothing about the history of mathematics, there is a very good chance you have heard of Pythagoras. He is probably the best-known mathematician outside of mathematical circles – but what do we actually know about him? We know that he was born on the Greek island of Samos around 570 BCE, and died c.495 BCE in ancient Kroton, Magna Graecia (now southern Italy). In his youth Pythagoras travelled widely, to Egypt and other parts of Africa, to Babylonia, and possibly as far as India. He would have learnt something from each place he visited, but we are not certain what exactly.

It was in Kroton, where he lived from around 530 BCE, that he founded an extraordinary – and probably the only – mathematical sect in the history of humankind. Its members not only learnt and taught mathematics to each other, but lived together in a kind of commune. We know some of their names: Hippasus (c.530–450 BCE), Philolaus (c.470–385 BCE) and Archytas (c.435–c.360 BCE). This group also included women mathematicians, among them

Pythagoras' own wife, Theano, and their daughter, Myia. More recently, historians have uncovered old sources that name no less than seventeen women Pythagoreans. These would have been called out because they were the most important in terms of their contribution to the group – in other words, there were almost certainly more women in the sect. So here, for the first time, we can name the first women mathematicians, although we can't say exactly what their contributions were to the Pythagorean development of mathematics.

The community itself was interesting, and extremely secretive: its members were bound not to divulge their knowledge – some of which, as we shall see, was mathematical – beyond their circle. We know there were two groups of Pythagoreans, the first being *akousmatikoi* (the listeners); as their name suggests, they listened to the teachings of the group, and were focused on the religious and ritualistic aspects of Pythagoras' teachings. Sometimes it is said that they were an outer circle of Pythagoreans, coming to hear the discussions but not necessarily living within the Pythagorean commune. Then there were the *mathematikoi* (the learners), who lived together and concentrated on the more mathematical and scientific aspects of Pythagorean thought. They were described by earlier historians as following strict rules of life, for example stringent dietary regulations (mandatory vegetarianism and, some say, a prohibition against eating beans), rigorous rules of self-discipline, and the complete observation of religious doctrine. This included the belief in the immortality of the soul and its transmigration; in other words, the Pythagoreans believed that the soul would survive the death of one person to be reborn in another. Though it is not entirely clear, there was some kind of mathematical belief connected to that too, related to the souls perishing once all incarnations had been exhausted – when a limit to the infinite had been reached.

For the Pythagoreans, numbers were at the centre of their understanding of the universe. (There were differing opinions – not all of them believed everything that we mention here.) They would have written numbers by using letters from the Greek alphabet, as was the case in Greek culture at the time, but the Pythagoreans also seem to have used dots, known as *psiphi* (pebbles), and by their possible formation they also developed an understanding of what we now know as figurate numbers. For the Pythagoreans, these number formations gained meaning that was more mystical than mathematical. The triad, 3, was considered a 'perfect number', important as it stood for three things that together amounted to the goodness of a person: having self-discipline, being hard-working and enjoying good fortune. Ascribing beliefs to the meaning of numbers is called *numerology*. The Pythagoreans practised numerology, contemplating numbers for their insights into the universe, and connecting the properties of a number to the imagined power that it may have in relation to people or situations.

```
        •
      •   •
    •   •   •
  •   •   •   •
```

The number 10 was another important number for the Pythagoreans. It was connected to the first *tetraktys*, a triangular number of ten points, which could be constructed from the triad. Being made up of the first four integers added together $(1 + 2 + 3 + 4)$ and presented in an equilateral triangle, the *tetraktys* represented completeness, order, balance and the harmony of the universe. The Pythagoreans don't seem to have believed in a kind of geometer God who created a rational universe (as some later thinkers did, such as Johannes Kepler: Chapter 15), but they did see numbers as having somehow made the cosmos, as well as making that cosmos understandable to humans. In other words, numbers were a link between the mysteries of the universe and the human understanding of it.

The Pythagoreans saw the very heavens moving mathematically – and musically. For them, the planets danced to an exquisite symphony which could be charted using mathematics, with harmonies that had their basis in mathematical equations. For example, they found that if you have a string instrument, and on it two strings of the same length, when plucked these two strings would sound the same. If one string is twice as long as the other, in other words the ratio of their lengths is 2:1 (2 to 1), they sound in harmony, and will produce an octave. Similarly, a ratio 4:3 is a fourth and 3:2 a fifth, all very pleasant sounds. Pythagoras' follower Philolaus described the musical scale based on mathematical ratios in his *On Nature*, the first book in the Pythagorean tradition, which dealt with topics ranging from astronomy and cosmology to music. He set out a cosmological system with an eternal fire (not the Sun) in the centre of the universe, around which the Earth and the other planets revolved. For the time, this was extraordinary: it was not until the seventeenth century that the view that Earth was at the centre of the universe was overturned.

Today, Pythagoras is most famous for the theorem named after him, that the sum of the squares on the two smaller sides of a right-angled triangle are equal to the square on the largest side (the hypotenuse). We already saw in Chapter 2 that the Babylonians and Egyptians knew how to find three numbers (which we now call Pythagorean triples) that you could use to construct such a right-angled triangle, though they did not provide a proof of the relationship between these numbers. Pythagoras and his followers did. But whether he was the author of 'his' theorem is a matter of hot debate. Philosophers such as Plato and Aristotle, Pythagoras' near contemporaries, tellingly did not give him credit for it, though accounts after them did. For centuries, people took Pythagoras to be a magus as well as a master mathematician. More recent scholarship concedes that we have really no evidence that the theorem originated from either Pythagoras or his followers or students. But it remains significant that Plato and Aristotle credited the Pythagoreans with being the first to develop the science of mathematics.

Plato (c.428–c.348 BCE) was deeply influenced by the Pythagoreans, although he was never a member of the sect. He is particularly important for mathematics because of his *Theory of*

Forms (or *Theory of Ideas*). In this theory, he describes in detail the difference between the 'real' and the 'ideal'. For Plato, 'forms' are perfect, unchanging, whereas real, physical things are subject to time and change, and can never be perfect. He uses geometry to prove his point. Any geometrical object is perfect only when it is imagined – in reality, we can never produce a perfect square or a perfect cube. We can get close, but it will always contain imperfections, even at the most minuscule level.

In one of his works, *Timaeus*, Plato paints a picture of the creation of the universe. He describes how the demiurge, a kind of primal force and divine craftsman, imposed mathematical structure on chaos to create our good and beautifully ordered cosmos. In Plato's account, fire, air, water and earth inhabit four regular solids, which we now call Platonic solids: taken in turn, the tetrahedron, the octahedron, the icosahedron and the cube. What is so special about these shapes? Well, they can all be made up of two types of right-angled triangles – in Plato's eyes, triangles that were better than all others. The first of the most perfect triangles is the one you get if you halve a square across its diagonal:

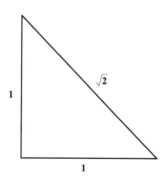

The second-best triangle, according to Plato, is the one which, if reflected across one of its sides, makes an equilateral triangle (where all three sides are equal):

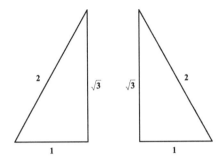

These two triangles (those whose sides are 1: 1: $\sqrt{2}$ and 1: $\sqrt{3}$: 2) can be used to construct the four solids: tetrahedron (four equilateral triangles), octahedron (eight equilateral triangles), icosahedron (twenty equilateral triangles) and cube (twelve of the best kind of triangle, doubled on their hypotenuse to make six squares).

They are four of the five regular solids you can make in three dimensions, the other being the dodecahedron, the fifth Platonic solid, the construction of which Plato does not discuss at length, but which is made of twelve regular pentagons, and for Plato represented the divine spark. These five solids are very special. They are the only regular solids in three dimensions made up of identical faces (regular polygons) meeting at the same three-dimensional angles.

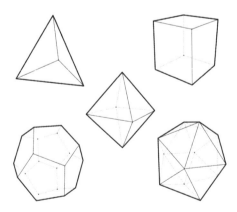

But there was one Pythagorean who paid a heavy price for divulging a secret contained in two of these solids and more broadly in the mathematics his sect was exploring at the time. This is the concept of *incommensurability*, now said to have been discovered by the Pythagoreans. Incommensurability is when the ratio of two magnitudes cannot be given in the same unit of measure. For example, the ratio between the side of a square and its diagonal are two such magnitudes: applying Pythagoras' theorem, the length of the diagonal would be $\sqrt{2}$ times the length of the side of the square, and as $\sqrt{2}$ is not a rational number, neither will this diagonal be a rational number. In other words, it will never be a whole number or even a fractional multiple of the side's unit of measure. In other words, these two lengths are *incommensurate* – you can't compare them with the same unit of measure. This very incommensurability can be seen in the Platonic solids. It is the ratio of the side and hypotenuse in Plato's first, best triangle. The side and diagonal of a pentagon, used to make the dodecahedron, are also incommensurate.

Legend has it that one or other of these incommensurable magnitudes was divulged by one Hippasus, a Pythagorean who came from Metapontum, to the uninitiated – people outside of the commune. We don't know if he discovered this concept or learnt it from other Pythagoreans. Whatever the case, what the Pythagoreans did in practice was considered protected knowledge, and spreading knowledge of incommensurability beyond the hallowed group of *mathematikoi* was an egregious sin. Sometime around 450 BCE, on board a ship off the south coast of Italy, Hippasus was judged to be a traitor of this mathematical community and was thrown overboard to his death in the raging sea. Why the secret was so jealously guarded is anyone's guess. Perhaps incommensurability threw into question the Pythagoreans' belief that everything could be expressed in terms of whole numbers or their ratios. No one knows for certain.

The Pythagoreans probably influenced more philosophers and mathematicians than any other group of mathematicians until the twentieth century (Chapter 33). For instance, as we shall see, the concept of incommensurability became central to the Greeks' adoption of geometry, where segments, areas and volumes were described through comparisons and ratios. Hippasus might have paid with his life, but the findings of this small, secretive mathematical sect invigorated the pursuit of mathematics for millennia to come.

CHAPTER 5

The Greatest Mathematical Bestseller of All Time

The Hellenistic world's very best scholars, most interesting objects and, apparently, every book then in existence, could be found at the majestic Mouseion (Museum) in Alexandria, on the Egyptian shores of the Mediterranean. Established by Alexander the Great's successor Ptolemy I and his son around the third century BCE, it gathered together the very best minds to preserve and further learning in many different subjects.

It's said that, not long after becoming king of Egypt, Ptolemy I announced that he wished to learn mathematics. He wasn't satisfied with the texts available; there were too many ideas scattered on different papyri, and the mathematical topics in them seemed unconnected and not very interesting. Were there any others that could teach him mathematics in an easy and engaging way? The learned men of Alexandria set about trying find a solution to satisfy their ruler. How could they organise all the existing knowledge of mathematics, scattered as it was across different books all over their great kingdom? Then someone had a brainwave: a mathematician named Euclid was famed as a brilliant teacher, and he was duly brought to Alexandria to write the book.

We don't know much about the man himself. Even the years of his birth and death are not certain; we think he lived between about 325–320 BCE to around 280–265 BCE. The mathematical philosopher Proclus, who lived around seven centuries after Euclid, said that he was a pupil of Plato, but the dates don't quite tally up. In any case, Euclid set out on the mammoth task of revising and simplifying all the mathematics known at the time and putting it together in one work. This he completed around the turn of the second century BCE, dividing it into thirteen books and calling the collection the *Elements* (*Stoikheîa*). It remains the most widely translated and distributed mathematical textbook ever created, endlessly copied, commented upon, translated, printed and produced in all kinds of formats for different audiences around the world.

Euclid didn't invent all the mathematics in the *Elements*. He compiled what was already known and organised it. He probably weeded out all that he found unimportant or which did not fit in the overall structure of the work. (He certainly knew more mathematics than is contained here because he wrote other mathematical treatises, such as one on optics, the geometry of vision.) It's clear that in the *Elements* he wanted to keep things as simple and neat as possible.

Nevertheless, for someone who's never studied mathematics before, the *Elements* is a pretty complex work. Ptolemy I found it so, and apparently asked Euclid, whom he had recently appointed to be the first teacher of geometry at the Mouseion, to make it easier for him to understand. In reply, Euclid famously quipped, 'There is no royal road to geometry'. There are no shortcuts, even for kings.

So what do we find as we embark on the journey? The *Elements* proceeds by *deductive reasoning* – that is, drawing valid inferences (conclusions from simple premises). In this way the work is structured so that statements of general principles and basic, self-demonstrative facts are progressively built on to explain ever more complicated problems. This method of working from the simplest statements to very complex conclusions withstood the test of time for nearly two millennia, and is something mathematicians still use today (although as we shall see in Chapter 23, one ambiguity was discovered).

Euclid was a master builder of intellectual structures, but he didn't invent this system of deductive reasoning. Long before him, Thales of Miletus (624–547 BCE) had already made geometry into a deductive science, and had derived some beautifully clear theorems which appear in the *Elements*. For example, the famous theorem that bears Thales' name states that a triangle inscribed in a semicircle, having its longest side as the diameter of a circle, is always right-angled.

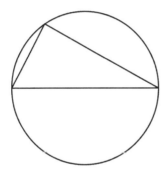

This is given in Euclid's *Elements*, in Book III, as Proposition 31. What are these propositions? We have already said that there are thirteen books in the *Elements*, and the work is structured in such a way that each book leads to the next – what was proved in one is built upon subsequently. Each book also proceeds deductively, starting with *definitions*, followed by *postulates, common notions* and finally *propositions*. In total, Euclid's *Elements* contains 131 definitions, five postulates, five common notions and 465 propositions.

Let's have a look at these more closely. The first of twenty-three definitions in Book I states that 'A *point* is that which has no part'. Definitions are there to establish the basic statements that we will take for granted from there on. Then come the postulates, of which there are only five in Book I. A postulate is a basic statement that should not be questionable in any way. The first postulate here says that it is possible to 'draw a straight line from any point to any point'. No arguing with that, but it needs to be stated for it to be used later. Then come common notions, again five of these in Book I. Common notions, also sometimes called *axioms*, refer to magnitudes and establish relationships between the elements (point, line, circle, angles). The first in Book I states that 'things which equal the same thing also equal one another'. And finally, we come to propositions, sometimes also called theorems. The first of the forty-eight in Book I says that it is possible 'to construct an equilateral triangle on a given finite straight line'. The propositions are about what could be *proved* or *constructed*. To prove something, in mathematics, is to show that it is true, beyond *any* doubt. To construct something in Euclidean geometry, you can use only a straight edge and a compass – the tools of geometry available at that time. These are called *Euclidean tools*. No measurements, no units of measure. Only the opening of a compass, a straight line, and comparison between magnitudes, whatever they may be (lengths, areas, volumes or angles).

We have already mentioned Thales, one of the Seven Sages of ancient Greece, a legendary group of thinkers (Thales was the only mathematician) which existed in the sixth century BCE. A maxim attributed to him, or at least to this group, is 'Know thyself', words that were engraved on the entrance to the Temple of Apollo at Delphi, believed to be the centre of the universe in Greek mythology. The history of mathematics is sometimes close to myth, and in ancient Greece, nowhere has it come closer than in this temple, in which a question was posed that kept mathematicians busy for the following 2,000 years.

Alarmed by the plague which was cutting a swathe through the Greek territories in the fifth century BCE, the inhabitants of the island of Delos became convinced that it had been sent by Apollo, god of the Sun (and much else), as punishment. They went to the Temple of Apollo to ask the high priestess Pythia, known as the

Oracle of Delphi, what they could do. The answer came back that in order to assuage the god they should double the size of Apollo's altar, an ornate ten-foot-high cube. That didn't sound very difficult to do. Double the cube? How hard could it be? The Delians readily agreed. But they were flummoxed. Double the length of the sides and you end up with a cube eight times the volume of the original. Change the height of one or two lengths and you change the shape. They even reached out to Plato, who was similarly stumped. This problem, now known as the *Delian problem*, rested on how to find the cube root of 2, and was eventually proven – not until the nineteenth century – to be an impossible task using only those Euclidean tools of geometry available in the fifth century (a straight edge and compass). As the clever Plato concluded, the dastardly god would rather the Greeks were busy thinking than living lives of wickedness and war.

The Delian problem was one of what are now called the three impossible problems of Greek antiquity. The other two problems were trisecting an angle (splitting it into three) and squaring the circle (constructing a square with the same area as a circle) – as always, only using the Euclidean tools. Some particular angles, for example a 90-degree angle, can be easily trisected (into three angles of 30 degrees), but it was finally proved two millennia later that it is impossible to do this for arbitrary angles. It was a similar story with squaring the circle. The Greeks knew how to calculate the area of a circle, for which a special number, Pi or π, is needed. Archimedes (287–212 BCE) is credited with approximating the value of π to between $3\frac{1}{7}$ and $3\frac{10}{71}$. This is remarkably accurate: we know π as 3.14159 (to five decimal places), and Archimedes' calculations put it between 3.14085 and 3.14286 (also to five decimal places). π is the ratio between the diameter and circumference of a circle (it is the circumference divided by the diameter), and the area of a circle is π multiplied by the radius squared. Some attempts to solve the squaring of the circle came close to succeeding. Hippocrates of Chios (c.470–c.410 BCE), whose own *Elements* served as a model for Euclid's later work, tried to fit some circular areas into rectangular ones and managed to do it, up to a point. He demonstrated that lunes (circular areas bounded by parts of two circles) can be measured in the same way as rectilinear figures. But in the nineteenth century, the problem of squaring the circle too was shown to be unsolvable using only a straight edge and compass.

For the best part of 2,000 years, mathematicians just couldn't leave these three famous problems alone. The question of how to solve them, and whether they even could be solved, was asked time and again. This shows us both that mathematicians don't give up easily, and that they don't mind passing on unsolved problems to the next generation. Someone, somewhere, will find an answer – or prove that one can't be found. This has been a constant characteristic of mathematics from its earliest days: a relentless pursuit of solutions across time and space, not (always) jealously guarded by the one who first formulated the problem. Who can resist the allure of a mathematical mystery?

Collectively, the ancient Greeks left a huge legacy for the world in their advancement of mathematics. Perhaps no other culture has done more overall in developing ways to structure mathematical knowledge, to use proof and to formulate a problem. Without doubt, no other book in any culture so boldly represents a unity of mathematical ideas as the *Elements* of Euclid. He gave us a near perfect structure, an intellectual edifice of abstract thinking that lasted for two millennia, a mathematical textbook that still has much to contribute to all who are learning our subject.

CHAPTER 6

Where's the Proof?

Chinese legend tells how Huang Di, the so-called Yellow Emperor and divine progenitor of the Chinese people, commanded one Li Shou to invent arithmetic in the twenty-sixth century BCE. This was alongside a directive to another of his ministers to come up with China's first writing system, so it goes some way to showing the longevity of China's interest in mathematics as well as the high esteem in which mathematics has been held throughout Chinese history.

The Later Han Dynasty (25–220 CE) was the period in which the boundaries that more or less define the nation of China as we know it today were established. It was also when *Jiǔzhāng suànshù* (*The Nine Chapters on the Mathematical Art*) first appeared. This remains the most famous mathematical treatise emanating from China. Western scholars compare it with Euclid's *Elements*, but this was a very different, original work. In particular, it was focused on problems, the proofs of which are not at all the same as those of the *Elements*.

The contents of the *Nine Chapters* were compiled over centuries, with the book itself assembled probably sometime in the second or first century BCE. The mathematician Líu Huī (c.225–295 CE), who wrote a commentary on this work in 263, said that the original was much older. The treatise drew on ancient texts, such as the *Suàn shù shū* (*Writings on Reckoning*), from a tomb sealed in the second century BCE, but the *Nine Chapters* was much better organised, more systematic, and the range of problems was much greater. It somehow survived the educational reforms of Qín Shǐ Huáng, founder of the Qin dynasty, who ordered the vast majority of the existing old books to be destroyed around 213 BCE, including all of those related to mathematics. Somehow, the contents were saved and reconstructed around 170 CE.

The *Nine Chapters* contains 246 problems, the methods for solving which could be used to help with practical issues such as taxation, engineering and surveying. The topics are organised and grouped, with each chapter generally taking the name of the first problem presented in it. The overall method is named and explained and then shown how it can be applied to a particular set of problems. Thus, Chapter 1 is called *Fāng Tián* ('rectangular field') and it is about problems to do with land surveying (areas and fractions). The method by which the problems will be addressed is called *fāng tián*. The first problem states: 'Now there is a field, width 15 *bù* [units] and length 16 *bù*. Find the area of the field. Answer says 1 *mǔ*.' (We deduce that 1 *mu* must therefore be equal to 240 square *bù*, and the way we have arrived at that problem is to multiply the values for the width and the length, but that is not stated in the book.) The book proceeds in a similar fashion. It sets out the procedure and gives it a name, states a problem and a solution, then that name of the procedure is recalled as the way to solve similar problems later on.

The great importance of that general-to-particular method in Chinese mathematics becomes apparent in another mathematical astronomy book from around the same time, called *Zhōubì suànjīng* (*Arithmetical Classic of the Gnomon and the Circular Paths of Heaven*): 'The method of calculation is very simple to explain . . . a person gains knowledge by analogy; that is, after understanding a particular line of argument they can infer various kinds of similar reasoning ...' This is somewhat different to Euclid's (Greek) particular-to-general method. The Greek method was based on proving a theorem from axioms, which could then be used to prove another theorem, and giving rigorous proofs of those results. The Chinese method was to show how a problem was solved, to give that type of solution a name, and recall the method by its name to solve further problems. For a long time, historians so attuned to Euclid's way of working did not recognise that there were proofs in the *Nine Chapters* and believed that Chinese mathematicians simply stated formulae without backing them up. But more recently we've appreciated that the Chinese did give convincing accounts of their methods for solving problems, not through proofs, like the Greeks, but through examples. It is no worse or better than the Greek method, just different.

Let's look more closely at that way of showing how things work out. Another problem from the *Nine Chapters* concerns the areas of rectangles *X* and *Y*. We want to show that these areas are equal to each other when the corners touch on the diagonal of the rectangle enclosing them. (This is actually the case *only* when they touch on the diagonal, but the proof in the *Nine Chapters* doesn't go as far as to say that.) As we can see, the enclosing rectangle is divided into two equal triangles, *ABC* and *ACD*.

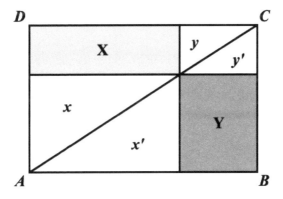

The areas of X and Y are equal to each other because the areas of the triangles are equal ($ABC = ACD$) and the remaining areas are equal (we can easily see that $x = x'$ and $y = y'$), so it must be that the area of X will be equal to the area of Y. But – and this is a huge 'but' in understanding how things are given in the *Nine Chapters* – this is not actually spelt out. The book does not use the sort of deductive reasoning that we ourselves have just used! In general, in this and other Chinese texts from the period, it is not shown how things are proved: the text doesn't show the steps, it simply states that this is so. In the Chinese tradition of mathematics from this period, only the final results are presented, and that is taken and applied in later problems.

This tradition of giving the final result at the end of stating a problem is perhaps related to the way things were formulated and written in Chinese script, and is noticeable in their way of working with numbers too. Let us look at the symbols for digits and numbers. From around the fourth century BCE digits were represented by small bamboo sticks which could be placed in a counting board.

This system was, like ours, a value notation system, which means that each digit had a meaning attached to it according to where it was placed – 1 in the place of units (on the right-hand side) is quite a lot different to the 1 in the place of hundreds (two places along to the left). The signs for digits of the first nine values were:

but only when they referred to units, hundreds, ten thousands, and so on. Note that I didn't include tens and thousands there: for those alternate places the signs were rotated to appear like this:

A given number would be written in a horizontal line from left to right in descending order, just like we do now. When there was no value (we would write 0 there), the space was left empty. So 4,506 would be written as follows:

The ancient Chinese performed computation using these rod numerals, but they would only note down the final result, and that number would be written out in words. In so doing, the actual method of computing was lost. Working in this way meant that learning mathematics from books was done much more by rote rather than through worked examples. Having a skilled teacher to guide you in how to understand the process was therefore even more important. But the beauty of this way of working was that Chinese mathematical texts from this period are concise, precise, and usually contain much more information than is immediately obvious. We can imagine ancient sages opening up the mysteries of those abstruse symbols and diagrams in their explanations to wondering mathematical students. Chinese mathematics is like that – it contains multitudes not necessarily visible at first sight.

Liú Huī presented in his commentary on the *Nine Chapters* a great many other mathematical methods. Perhaps the most impressive was his algorithm for solving three simultaneous linear equations. An *equation* is a mathematical statement that contains some known quantities given by numbers, and unknown quantities given by letters. Solving an equation means that we find the value for the letter of the unknown quantity. You can solve a single *linear* equation – when the power of the unknown quantity is 1 – with one unknown quantity. If you have two unknown quantities in a linear equation, you need two equations to find values for both unknowns. For three unknown quantities, you need three linear equations – and so on.

Líu also made a major contribution to Chinese mathematics with his *Hǎidǎo suànjīng* (*The Sea Island Mathematical Manual*), an extended appendix to Chapter 9 of the *Nine Chapters*. There he presented a theorem called *Gōu-Gǔ dìng lǐ*, known in the West as Pythagoras' theorem. He showed how it could be used to calculate the heights of objects one couldn't otherwise measure directly, if they were very high or distant, for instance. Líu's presentation is particularly beautiful; he explained how things work by colouring the areas of squares, and then the translation of the areas, to show that the algorithm works.

There were other significant Chinese contributions from this time. Sūn Zi (c.400–460) wrote *Sūnzi suànjīng* (*Master Sun's Mathematical Manual*) where he showed how what is now called the *Chinese remainder theorem* works. This theorem starts from an unknown number. Given the remainders of its division by different numbers, by constructing an equation you will be able to find the unknown quantity. It is amazingly useful as it also provides the possibility of finding any other bigger number which would also be divisible by the divisors originally used. These multiples of the original unknown quantity would also have the same remainders when divided by the same numbers. This theorem and procedure have had many applications in the centuries that followed. Sixth-century Chinese astronomers used it to measure planetary movements. Today, it finds use in computer science, particularly in relation to generating keys to encrypt and decrypt data.

In the early seventh century, Emperor Táng Gāozǔ issued a decree demanding a reform of all education. But unlike his predecessor Qín Shǐ Huáng, Gāozǔ didn't want to obliterate mathematical knowledge. On the contrary, he sought the very best texts that scholars could study to learn all the important subjects. It sounds a little like Ptolemy I's earlier initiatives, which resulted in Euclid's *Elements*. Three scholars heeded the emperor's call, selecting, annotating and sometimes correcting the most important mathematical works then in existence, which were then used at the national university. What became known as the Ten Mathematical Classics (although there were more than ten) included some of the works we've seen here, such as the *Nine Chapters*, the *Sea Island Mathematical Manual* and *Master Zi's Mathematical Manual*. The teaching of mathematics finally became formalised, one strand of an educational drive that aimed to unify a country divided for three centuries.

Mathematics in the two geographically large but distant regions of China and the western, Mediterranean basin, developed more or less independently. Knowledge of each other's cultures travelled along the trade routes, of course, but gaining a clear understanding of the ways things were done mathematically was not always an easy task. We have seen how different the ancient Greeks and ancient Chinese were in the ways they communicated and proved their mathematics, but the mathematical truths still remained very much the same.

CHAPTER 7

The Dawn of Number Theory

'Here lies Diophantus, the wonder behold.' So begins a curious riddle about the greatest mathematician of late antiquity. He was a native of Alexandria, that beautiful Greek city nestled in the Nile's delta, and lived in the third century. What was his life like and how long did he live for? The riddle tells us, and the means by which it does points to why Diophantus was so important to mathematics.

Let's look at this fun little epigram, penned by the grammarian Metrodorus three centuries after Diophantus died:

God gave him his boyhood one-sixth of his life,
One twelfth more as youth while whiskers grew rife;
And then yet one-seventh ere marriage begun;
In five years there came a bouncing new son.
Alas, the dear child of master and sage,
After attaining half the measure of his father's life chill fate took him.
After consoling his fate by the science of numbers for four years,
   he ended his life.

So we have Diophantus's boyhood lasting a sixth of his life; after a seventh more his marriage taking place; his beard growing after a twelfth more; his son being born five years after that; his son living to half his father's age; and Diophantus dying four years after his son.

We can use this information to compose an equation. The way we write it today is very different to how it would have appeared back then, of course: apart from the numerals, which we write quite differently now (see Chapter 10), the Greeks lacked the equals sign (which was invented only in the fifteenth century, as we shall see). Let's say that the unknown quantity, Diophantus' life span, is x. Then, our equation would look like this:

$$\frac{1}{6}x + \frac{1}{12}x + \frac{1}{7}x + 5 + \frac{1}{2}x + 4 = x$$

By solving this equation, we find out that Diophantus lived for eighty-four years, from around 200 to 284. We discover a few things about his private life, his personal triumphs and tragedies, even his emotions (if true). And, probably most importantly, it shows us one of the very things that he was most famous for – making up equations just like this one.

In the history of mathematics, Diophantus is best known for his famous work *Arithmetica* and his treatise *On Polygonal Numbers*. *Arithmetica* is the earliest known work that presents ways of solving arithmetic problems using algebra. Mathematical historians generally agree that *Arithmetica*, like Euclid's *Elements*, was a collection of mathematical knowledge of the time – and, as we shall see, Diophantus himself did not invent the method of algebra. Nevertheless, *Arithmetica* was so important it continued to inspire mathematicians for many centuries, remaining relevant even into our lifetimes.

Originally written in thirteen books, only the first six have survived. In them we can see Diophantus' rather unique method of showing how things work: he does not give a general method but reveals by example.

He recognised that not only whole numbers, which we call *integers*, but also all rational numbers can be used in equations. Rational numbers can be represented as ratios of integers; they are fractions of two whole numbers. We saw a bunch of those in the equation we constructed from the riddle about him. Now, however, *Diophantine equations* usually refer to those that have integer *coefficients* (the numbers multiplying the unknown quantity – for example, above, the numbers before the xs).

Diophantus worked on both *determinate* and *indeterminate* equations. Determinate equations are those that have an exact number of solutions and indeterminate are ones that don't. You will recall from Chapter 6 that if you have more than one unknown quantity in linear equations (of power 1) then you need to have the same number of equations as there are unknown quantities to find a solution, otherwise the set of equations will be indeterminate. But you can also have indeterminate equations of other powers (for example, of power 2, which we call *square* or *quadratic* equations).

Mathematicians still mull over some of the problems Diophantus gave us. In *Arithmetica* he references propositions from a now-lost text entitled *Porisms*, the author of which is unknown. These porisms, which are very much written as statements, are not proven. One example of a Diophantus porism is the statement that 'the difference of any two cubes is also the sum of two cubes'. This means if you have some numbers, say $a^3 - b^3 = c$, there will be some other numbers that will satisfy $c = p^3 + q^3$. The trick here is to find such rational numbers (they don't have to be integers). This looks simpler than it is! Another porism is that 'any square number can be resolved or given as a sum of two squares in any number of ways'. We can try that with 25, which is 5^2. From those right-angled triangles we saw earlier, we can also say that $3^2 + 4^2 = 5^2$. But *any* square number? The porisms have continuously tantalised and inspired later mathematicians. They were particularly important for the development of number theory, a branch of mathematics primarily concerned with positive whole numbers. Because this was so different from the mathematics of Euclid and Archimedes, it is often thought that it emanated from non-Greek sources. Alas, we don't have enough information to say either way.

The other important and, by all accounts, rather beautiful treatise was Diophantus' book *On Polygonal Numbers*, of which only a fragment survives. It was very different from *Arithmetica* as it offered strictly geometric proofs to propositions. A *polygonal* number is one that can be arranged as a collection of dots in a shape of a polygon (only regular polygons feature here). We saw earlier that the Pythagoreans already used dots to represent numbers (Chapter 4), but now Diophantus looked further. For example, triangular numbers would be 1, 3, 6, 10:

Square numbers (1, 4, 9, 16, 25 . . .) follow the shape of a square, pentagonal numbers (1, 5, 12, 22, 35) the shape of a regular pentagon, and so on. Diophantus produced formulae to find various properties of polygonal numbers, so that you could always find a particular *n*th number of that sequence.

As with the porisms, here we think Diophantus was summarising mathematics that was known already rather than coming up with the idea himself, although he did it in a rather nice and precise way. He said that he learnt this particular formula from his predecessor, the Greek mathematician Hypsicles (190–120 BCE). Nevertheless, we see it for the first time in Diophantus' work, and even if he stood on the shoulders of giants, we owe him a debt of gratitude. Through his work earlier mathematics was preserved and bettered. It is not only the mathematicians who originate new mathematics who are important; we owe it to those who comment and build on the mathematics they inherit to achieve further developments.

One important commentator on Diophantus was Hypatia (c.350/370–415). She is the first woman mathematician whose work we know. Unlike with the Pythagorean women, we can be certain of the individual contribution she made to the field. Her sex alone would make her stand out in the history of mathematics, rare as women are, especially from this period, and she is still celebrated the world over for her involvement with mathematics.

The daughter of a mathematician, Theon of Alexandria (335–405), Hypatia was famous in her own right as a mathematician, astronomer and philosopher. Jointly with her father, she composed a commentary on Claudius Ptolemy (c.100–170), whose work made sacrosanct the geocentric model of the universe with Earth in the centre, and the Sun and all the other known planets in the solar system and beyond circling around it. Despite the fact that this was in itself wrong, Ptolemy's *Syntaxis Mathematica* was important and developed a number of theories useful for the development of both mathematics and astronomy. What her own contribution was to the commentary is hard to glean, however.

We do know that Hypatia developed an approximate method for doubling the cube as well as calculating cube roots – which was needed to solve the impossible Delian problem we looked at in Chapter 5. (To end up with a cube with a volume of 2 units cubed, you would need sides of length $\sqrt[3]{2}$, as that side, when cubed, would give 2: $\sqrt[3]{2} \times \sqrt[3]{2} \times \sqrt[3]{2} = 2$). She also produced commentaries of Euclid's *Optics* and the *Conics* of Apollonius, to make them easier to understand. Hypatia taught in the school her father founded on the model of the old Mouseion; the great Alexandrian library was by that time already destroyed. She taught philosophy, but we don't know whether she taught mathematics alongside it. Hypatia's philosophy was neo-Platonic: it revived Platonism. In this school of thought, the universe is derived from a divine principle, but continues and will continue forever without end.

Hypatia became a famous philosopher in her time and one of her students, Synesius of Cyrene, became bishop of Ptolemais. Quite a lot of what we know about her we gather from the correspondence between the two. He even asked for her help in constructing mathematical and astronomical instruments.

Another of her famous students was Orestes, the Roman prefect of Alexandria – a pagan, like Hypatia. When Cyril became patriarch of Alexandria in 412, tensions flared and riots broke out between Christians and non-Christians, particularly Jews. Hypatia became a focus of their anger, and she was killed.

Hypatia became a symbol of many different causes after her death, so we shouldn't be surprised to find her name linked to companies in the fields of technology, software engineering, education, and supporting women in leadership. She also features in films and novels. Her greatest contribution, however, has been to preserve, through her commentary, the work of Diophantus and the first six books of his *Arithmetica*.

Mathematics continued to thrive in this period north of Alexandria, in the Bosphorus. Constantinople (today's Istanbul) was the capital of the Byzantine Empire. Two important geometers are known from it, Anthemius of Tralles (c.474–c.534) and Isidore of Miletus (c.475–?). The two men were capable mathematicians. Anthemius' best-known contribution is the construction of an ellipse with a string fixed at the two foci, and his work on the focal properties of the parabola. Isidore produced an important compilation of Archimedes' work and studied the properties of vaulted roofs. These were useful skills. Around 532 the Byzantine emperor, Justinian I, brought them to the city to lend their mathematical nous to the designing and building of a splendid new Christian cathedral, the Hagia Sophia. The two men developed new mathematics that would allow the construction of the greatest dome ever built, and one of the first of its kind (a circular form sitting atop a square room). It was the crowning achievement of one of the world's most lavish and much-copied monuments.

Like many of the mathematicians we've seen in this chapter, Anthemius and Isidore also preserved some work from Greek antiquity, most importantly the work of Archimedes. Slowly, however, the focus was shifting, and the destruction of Greek and later Roman civilisations meant that for many centuries Greek mathematical knowledge would be lost to the rest of Europe.

CHAPTER 8

The Origin of Nothing

Since numbers were first invented and used, there has been an understanding that somewhere amongst them was their opposite – a nothing, an emptiness, a lack of quantity. Now we call it zero, and we have a symbol: 0. But it wasn't always thought of as a number itself – the concept existed long before the name and the sign. It was in first-millennium India that this nothing became formalised as a mathematical something.

Indian culture has a long history of mathematics. The sixth-century *Śulbasūtras* are geometrical texts that detail the proper arrangement of bricks to construct fire altars used in religious rituals that went all the way back to 1500 BCE. Pythagoras' theorem appears here (although of course it was not named after that Greek mathematician), but zero does not. Other cultures, such as the Maya, Mesopotamians and Greeks, had ways of dealing with zero but not by using a numeral for it. Instead they used a placeholder, leaving a space where the zero would go. Some concept of zero was required by the place-value systems used by the Mesopotamian empires (as we saw earlier, sexagesimal: a system of base 60) and the Maya and Aztec cultures (vigesimal: a system of base 20). So although it was acknowledged that there should be some way of dealing with zero, in these instances, zero did not get a full-fledged treatment as a number in its own right.

Several things point to India as being the place that birthed the numeral zero as we know it, or close to it, in the first millennium. A symbol, *śūnya*, which stands for empty, void or vacant, had a purpose in calculations in Sanskrit's word-numeral system, some of which have been found before the Common Era. Nothingness has deep roots in the philosophy and religions of India, which may have encouraged the development of a symbol for it. There is a belief that the sign of zero was linked to a particular meditative practice; in Hinduism and Buddhism, a circle with a dot in the centre symbolises a meditation on concentrating on the nothingness within ourselves, the empty space. That is not very far away from the 0 we now recognise. For a time it was thought that the world's oldest zero could be found in the ninth-century Chaturbhuj Temple in Gwalior, central India, carved into solid rock. On one of the temple's walls, the number 270 is written with a dot in the place of where now we would write 0.

We don't know who inscribed that particular mark, but we do know the names and work of three first-millennium Indian mathematicians who are credited with the invention of zero, and contributed much else to the development of mathematics.

Āryabhaṭa (476–550) is often said to be the first mathematician to use 0. He is also the earliest Indian mathematician to be known by an individual name; until him, mathematical works were attributed to collective or symbolic names. Even today he is celebrated as an icon of Indian mathematics. His work influenced many other famous mathematicians, some of whom we will meet shortly.

As with the *Śulbasūtras*, the mathematics Āryabhaṭa wrote was in verse. This was an established practice in India. It meant that problems could be more easily learnt and remembered through repetition, although it also gave rise to the possibility that students would repeat things they didn't necessarily understand – a well-versed tutor would still be important. Today, such verses or funny phrases are called mnemonics. If they are really good, they will be easy to remember, and can be helpful in recalling tricky operations in mathematics. One I like is 'Please excuse my dear Aunt Sally' for remembering the order of operations in an equation: p for parentheses (brackets), e for exponents (powers), m for multiplication, d for division, a for addition and s for subtraction.

Āryabhaṭa's most famous work, and the only one that has come down to us, was the *Āryabhaṭīyam*, an astronomical work in Sanskrit. Āryabhaṭa was interested in solving problems involving linear equations of the first degree, and constructing equations in a certain way. In particular, like Diophantus (Chapter 7), he was interested in indeterminate equations, where we can prescribe what possible values for the unknown quantity will satisfy the equation. Āryabhaṭa used only positive integers both for the coefficients and the unknown quantities.

One problem he described in some detail was about the value of pearls, where two men have different numbers of pearls and coins, but the total value of what they both have is equal. Āryabhaṭa's instruction to solve the problem was to divide the difference of the number of coins by the difference of the number of pearls, giving the price of one pearl. Āryabhaṭa gave solutions to the problems he stated, so that his readers could learn the procedure and apply what they learnt in other circumstances.

Most later mathematicians studied Āryabhaṭa's mathematics through the commentary on his most famous work by his successor, Bhāskara I (600–680), written around 629 and called *Āryabhaṭīya-Bhāṭya*. Bhāskara I was the second Indian mathematician attributed at one time or another with the invention of 0. Bhāskara I also built on Āryabhaṭa's indeterminate equations in his own work, introducing equations where both unknown quantities are squared (so, in the format $ax^2 - by^2 = c$). This was a little more complicated and a lot more interesting for mathematicians who came after him, in fact all the way into the seventeenth century: Pierre de Fermat, whom we'll meet in Chapter 18, used these in his own musings on numbers.

The final Indian inventor of 0, and the first to properly show how it could be used in calculations, is Brahmagupta (598–668). He wrote two important manuscripts in mathematics and astronomy. In one of these, *Brāhma-sphuṭa-siddhānta* (628), he gave some important descriptions of how zero should be used. He knew, for instance, that zero multiplied by any number gives zero. He didn't get everything right, though. He thought that you can also divide by zero – which you can't.

Let's see why not. Multiplication is like adding, just quicker. If you have 5 + 5 + 5 + 5 it is the same as 5 × 4 = 20. In the same way, subtraction is same as division, just quicker. It takes us four goes taking 5 away from 20 to get to 0: so 20 ÷ 5 = 4. Applying this same thinking, try to divide 20 by 0 to see how many times you do that to get to 0. Well, you can see straight away that 20–0=20, and we could do that forever without getting anywhere. You could then say that 20 ÷ 0 = ∞, the symbol for infinity. The trouble is that infinity is not a number. We can say that there are infinitely many things in something, or that something approaches infinity, but it is not a number. In other words, we can't divide by zero.

But Brahmagupta did do one very important new thing: he treated zero as a number, and he developed methods for dealing with zero in calculations. He also used the concept of *debt* which would, many centuries later, be introduced into mathematics and, centuries after that, be transformed into negative numbers (Chapter 13).

Recent research has uncovered yet more evidence about the origins of 0. A number of historians now ascribe the first appearance of 0 to the Bakhshálí manuscript, named after the village in which it was found in 1882 by a man digging for stones. Written on birch bark, it was so fragile that a large part of it was destroyed when it was first excavated. Its writer is not known. He signed himself son of someone called Chajaka and a 'king of calculators'. There are some seventy folios of mathematical problems with examples, all given in sutras and commentaries. Here, zero is assigned a digit, though it still doesn't look like the zero we write today: it is a dot. But it is treated as a number just like the other nine digits in our number system.

It also appears as an arithmetical operator – the author comments that adding a zero to a number leaves the number unchanged.

Radiocarbon dating by the Bodleian Library, where it is now held, and other research on the Bakhshálí manuscript has given wildly different results, dating it anywhere between the third and tenth century, although the writing on it is consistent and seems to have issued from a single hand. It shows similarities to Bhāskara I's work, although some of the contents could pre-date this. So we can't really say whether it was written before or after Brahmagupta's own discoveries, or whether the Gwalior temple can claim the first zero. Most historians now date the manuscript to between 650 and 900.

Such a simple sign, and such a complicated history, with claims and counter-claims for being the first to invent and use it. And no wonder – the creation of zero as a numeral was one of mathematics' most fundamental breakthroughs. It's an indispensable part of writing whole numbers, doing calculations and working in the decimal system which we use today. That little dot grew in size, developed an empty, hollow centre and travelled across the globe. Those ancient Indian mathematicians, known and unknown, planted and nurtured a seed that helped to create the modern world.

CHAPTER 9

Echoes from Baghdad

As wars and instability spread over Greece and its lands stretching across the Mediterranean, new cultures elsewhere were on the rise and would come to global prominence in its place. Baghdad became a centre of learning during the Islamic Golden Age, from the eighth to the fourteenth centuries. There, the House of Wisdom (Bayt al-Kikhmah), built by Caliph Harun al-Rashid in the late eighth century, held hundreds of thousands of books and objects of huge importance from the past. Scholars from all corners of the empire came to exchange ideas, talk to each other, translate old books – keeping alive much of the foundations of Western thought – and write new ones. In this place, mathematicians set to work on establishing a programme of collecting mathematical knowledge from around the world, translating it and updating it with their own inventions.

It was here that algebra was first named. It comes from the title of a book written by Muḥammad ibn Musā al-Khwārizmi (c.780–c.850), a polymath and head of the House of Wisdom from

820, who made huge contributions to mathematics, astronomy and geography. This book was *al-Kitāb al-Mukhtaṣar fī Ḥisāb al-Jabr wal-Muqābalah* (*The Compendious Book on Calculation by Completion and Balancing*), written sometime during the period 813–33. That *al-Jabr* in the title, or *algebra* as we now know it, comes from the Arabic for 'the reunion of broken parts', and it became a shorthand name for what is a system of techniques explained in that book. It refers to the balancing of algebraic equations as we move things about from one side to the other over – today – the equals sign, which is central to how algebra is done. Of course, when al-Khwārizmi was writing his book, the equals sign didn't exist, and neither did many of the symbols that we use today. Mathematical problems were described in words. Rather than *x*, for instance, meaning the unknown quantity, al-Khwārizmi used the word *shay* (Arabic for 'thing'). But he outlined the same procedure of keeping the two sides of an equation balanced, like a pair of scales. Anything added to or taken away from one side has to be done to the other. It is all about the art of balancing. Maybe those who find algebra a bit scary would be more inclined to try if they saw it this way.

Some of the most important contributions of the Islamic mathematicians lie in this area of mathematics. They took the understandings of the Babylonians (unsurprisingly, given Islamic mathematics grew in the same geographical region) and built on the classical Greek heritage of geometry, producing something new. They absorbed the idea that a mathematical problem could not be considered solved unless one could demonstrate that the solution was valid. The Greeks had already provided geometrical proofs, and Diophantus in particular had already begun work on providing principles to give a number as a solution to an equation. Now al-Khwārizmi stated that 'what people generally want in calculating ... is a number', that is, the solution of an equation. Algebraic manipulation could show that something is just so – a different proof to that which the Greeks had developed. The seeds of what now started to be considered and called algebra – the elusive art of balancing where algebraic manipulation would lead to a solution – sprouted in Baghdad's rich soil.

The quantities that al-Khwārizmi dealt with were generally of three kinds: a square (of an unknown quantity); a root of a square number (unknown quantity); and absolute numbers (the constants). Using these quantities, he listed six types of equations depending on what the squares, roots and numbers could be equal to, and suggested the ways to solve such problems.

He then proceeded to show how unknown quantities could be found using the newly discovered art of balancing. Al-Khwārizmi and other Islamic mathematicians of this period did not deal with negative numbers. In his system of algebra, coefficients, as well as the solutions to equations, had to be positive.

Al-Khwārizmi described a procedure of solving quadratic equations that has been given the name 'completing the square'. This procedure was already known in ancient Babylon (and can be found on a tablet at Yale known as YBC 6967). Let's take as an example the quadratic equation $x^2 + 10x = 39$. We can solve this by constructing the expression geometrically, as actual shapes. This would look like one square with sides x and one rectangle with sides x and 10. Halving the rectangle and moving it to the bottom of the square leaves a little bit missing in the corner – we complete the square by adding a small square with sides 5 (half the known width of the rectangle). This means we can simplify the equation: we now have a square with sides $x + 5$, not forgetting the little square with area 25 (5×5) that we have added: $(x + 5)^2 = 39 + 25 = 64$. A square which is equal to 64 has a side 8, so $x + 5 = 8$, and that means $x = 3$.

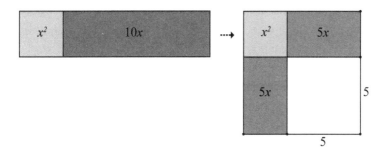

Is it easier than any other way? It depends on what kind of coefficients you have. This method is still used in many classrooms around the world today. And there is another thing, very great yet very basic, that we still do pretty much as al-Khwārizmi and his friends did back in Baghdad a thousand years ago: we use Hindu-Arabic numerals, the system with its digits 1, 2, 3, 4, 5, 6, 7, 8, 9 and of course 0. Al-Khwārizmi wrote a treatise on them, based on an Arabic translation of Brahmagupta, which hasn't survived (although some later translations have; indeed the Latin translations led to the mistaken belief that our system of numeration is Arabic in origin).

Al-Khwārizmi was by no means the only significant mathematician from this era and the House of Wisdom. Sometimes called the father of Arab philosophy, al-Kindi (801–73) was a prolific scholar who applied his mathematical mind to all sorts of matters. He worked on mathematics as it related to music, which was an essential part of Islamic culture. This wasn't entirely new: since the time of Pythagoras, music had been linked with mathematics. As any string instrument player knows, when you press a string with your finger, you divide it into a ratio. Some of these ratios would produce pleasant sounds when the string is vibrated, and some wouldn't. Al-Khwārizmi referred to rational and irrational numbers as *audible* and *inaudible*. The rational number ratios produce pleasant sounds; irrational ratios, also known as surds, unpleasant. Rational numbers are those that can be expressed as a ratio of two whole numbers. Surds have an *irrational* part; for example, $\frac{1}{\sqrt{2}}$ is a surd, as $\sqrt{2}$ is an irrational number.

Al-Kindi also wrote an important treatise in applied mathematics, developing the theory of cryptography, the science of ciphers, in the *Manuscript on Deciphering Cryptographic Messages*. This was the first known instance of statistical inference (drawing conclusions based on statistical analysis: that is, counting and classifying the information you have) and frequency analysis (working out how many times certain events occur). In cryptography this works on the principle that some letters are used more often than others in a spoken or written language, and from that information you can work out, with some certainty, the frequency by which they would appear in a text, replacing the ciphers with that letter and deciphering the coded message. The brilliant al-Kindi had such a command of linguistics, statistics and mathematics that he revolutionised the science long before the twentieth-century code-breakers at Bletchley Park.

Unfortunately not much of al-Kindi's writing has survived. Apparently he wrote some thirty-two books on mathematics, nine books on logic and many more on philosophy and theology. Remarkably, some of his works have been rediscovered as recently as the twentieth century, so perhaps more will come to light. But what has come down to us was hugely influential.

Al-Karaji (c.953–c.1029), who taught at the House of Wisdom, was another important mathematician who contributed especially to number theory. He lived in Baghdad but later moved to the mountains where he seems to have applied himself to living the good life and drilling wells. Al-Karaji introduced *inductive* argument through his study of arithmetic sequences. We saw deductive argumentation in the work of the Greek mathematicians: you start from a simple premise and make conclusions, and assuming the premises are true, the conclusion is correct. Inductive reasoning is different: here you set out to provide a universal truth through taking particular examples and generalising about them.

Al-Karaji proved some beautiful things with numbers. For instance, he looked at the sums of the first natural numbers, the squares of these numbers and their cubes. He proved that the sum of the first n natural numbers was $\frac{1}{2} n(n + 1)$. Try it and see: $1 + 2 + 3 + 4 = 2 \times 5 = 10$. One of the most pleasing results in number theory was his proof of the sum of integral cubes. He found that adding the cubes of the numbers that follow one another in their natural order is the same as multiplying their sum by itself. That doesn't sound at all likely, but it is true. For example, $1^3 + 2^3 = (1 + 2)^2 = 9$. Al-Karaji's inductive proof showed that this works up to the number 10 (he did not generalise it for n, standing for 'any number'). Neither did he work this out or prove it by drawing it out in overlapping squares (though that's one of the ways you can easily see how this works). In fact, al-Karaji didn't provide visual proofs at all – he was the first mathematician to break away from geometrical proofs and replace them with the type of algebraic ones which are at the core of algebra today. The algebra developed by Arabic mathematicians didn't illustrate formulae or provide solutions to equations in diagrams. It was all about the numbers.

Baghdad was a centre of learning for centuries. So many mathematicians can be traced to this fertile origin. But it was not only the remarkable mathematical discoveries made at this time and place that made it so important for the history of mathematics. The House of Wisdom preserved the ancient knowledge of the Greeks in a time of great migrations and wars in Europe. It was the mathematician al-Ḥajjāj ibn Yūsuf ibn Matar (783–833) who first translated Euclid's *Elements* into Arabic. And it was his translation that, as we shall see in the next chapter, was discovered by an English traveller, translated into Latin and brought back to Europe, restoring the hitherto lost Greek knowledge. Other scholars, too, drank deeply at the Baghdadi well; Gerard of Cremona translated into Latin al-Kindi's works and those of many other Arabic and ancient Greek thinkers in mathematics, astronomy, medicine and other sciences, thereby galvanising Western European thought.

Mathematics is always a voyage of discovery, but one which mostly starts from something known, even if the place you want to get to is not yet visible on the horizon. Rather than diving headfirst into uncharted waters, you may reveal some directions of travel from those who have gone before. The House of Wisdom provided superb opportunities for new mathematicians to do just that – to learn from the old masters and reveal the object they were trying to understand. And those mathematicians, in turn, revolutionised their subject, giving the world algebra, new proofs in number theory and the numerals we still use today.

CHAPTER 10

A Mathematical Tapestry

Every era weaves its own fabric of discoveries and ideas. Few were richer than that of medieval Europe as it slowly reawakened in the eleventh and twelfth centuries. There has been much debate about the so-called 'Dark Ages' – the period between the fall of the Western Roman Empire and the turn of the second millennium, marked, it has been said, by cultural and intellectual decline – arguing that they were not so dark at all. There were certainly times and places where learning survived and even thrived, but the continuous wars, falling populations, great migrations, religious orthodoxy and declines in trade all meant that European thought was overwhelmed. The old mathematics, like much of the intellectual contributions of the ancient world, was lost and, as we have seen, Arabic mathematicians took up the baton of preserving the heritage and pushing the field onward.

The medieval scholars who travelled to the Middle East and Mediterranean discovered many mathematical innovations. Algebra was one. The new numbers – Hindu-Arabic numerals, including 0 – another, serving as a shortcut to new ways of doing calculation; Roman numerals were until then still in use in the Latin West, and they made calculating slow and difficult. Then there was the ancient knowledge of geometry preserved by the Arabs through their translations and adaptations.

These were the most captivating things the travelling mathematicians came across, and they started looking at how they could bring them back to Europe and their own places of learning. New universities – starting with Bologna, Oxford, Cambridge, Salamanca and Padua – and schools attached to cathedrals were being founded all over Europe at that time. Universities were different from the ancient academies, such as those in Greece, or Baghdad's House of Wisdom. They represented a new concept in the learning world. Universities were given independence from the state and developed their own statutes and codes by which their members lived and worked.

But what to teach in such places? The first Latin translation of the central geometrical text, Euclid's *Elements*, was made not from Greek but from Arabic, by monk and philosopher Adelard of Bath (c.1080–c.1142/52). Adelard left England first for France, where he studied and taught, and then went to the sun-kissed shores of Italy and Sicily, which had been under Muslim rule for most of the ninth century. He travelled on through the lands of the Crusades as he called them: Greece, Anatolia, Syria (especially Antioch) and Asia Minor. He would have been immersed not only in Arabic traditions but the Arabic language, which was then still spoken even in Sicily, and he became fluent. He was fascinated by the Arabic works, especially in mathematics, astronomy and geometry, that he discovered on his long journey.

Adelard brought this learning, old and new, back to his own land. His travels had been paid for by King Henry I, and upon his return Adelard became tutor to his son, Prince Henry, the future Henry II. Adelard loved mathematics. He impressed upon his young scholar the value of the Greek learning and Arabic discoveries 'concerning the sphere, and the circles and the movements of the planets . . . Therefore I shall write in Latin what I learnt in

Arabic about the world and its parts.' Among these were the astronomical tables of al-Khwārizmi and books on algebra, as well as philosophical and scientific tracts, and a practical work on how to use astronomical instruments such as the astrolabe. And it was Adelard's translation of the *Elements* that established the Euclidean canon, with most future editions relying on his version of the work. It re-established the *Elements* as a central mathematical textbook that scholars in the Western world could draw upon, in Latin, the new language of learning.

In the dedication to his nephew of one of his translated works, Adelard said that he preferred to write books in which he would present the learning of others rather than his own, as the current generation wasn't that keen on modern inventions. Does that mean that Adelard himself felt he had something original to say? It is a tantalising prospect from a man so clearly capable and who loved mathematics.

By the end of his career, Arabic studies had been established in England, and a European intellectual renaissance was underway. A whole host of people would be inspired by the learning Adelard had begun to make available to them from faraway and ancient civilisations – not only in the contents of the books now accessible, but in the fact of translating them. These all formed part of a new great translation project in Western scholarship, bringing Greek works and Arabic knowledge to the Latin West.

It was by way of this that one of the most important mathematicians of this period became famous. Leonardo of Pisa, better known as Fibonacci (c.1170–c.1240/50), spent time in Bugia (now Algeria) with his merchant father as a young man, studied calculation with an Arab master, and later travelled to Greece, Egypt and the eastern Mediterranean, absorbing the different languages and mathematical systems there. Hindu-Arabic numerals were then known to a handful of European scholars from the writings of al-Khwārizmi, but it was Fibonacci's 1202 work, *Liber Abacci* (*Book of the Abacus*, or *The Book of Calculation*), which really brought them to the West's attention.

There, Fibonacci explained the methods that would enable people to put away the abacus and instead use the new numerals and ways of calculating with them. He described them in descending order – 9, 8, 7 and so on – giving 0 at the end, which he said the Arabs called *zephyr*. By these ten figures all together, Fibonacci said, any number whatsoever can be written. He gave a table to show the new numerals, showed how they should be used in calculations, and then gave some methods that one could apply, the *algorisms* (algorithms). Originally, 'algorism' simply referred to the Arabic system of numbers. Today, an algorithm signifies any predesigned procedure.

Fibonacci is probably best known not for introducing our modern system of numerals, but for revealing a beautiful number sequence which has taken his name. The Fibonacci sequence grows by adding the previous two terms to produce the next one: 1 + 1 = 2; 1 + 2 = 3; 2 + 3 = 5, and so on, giving the sequence 1, 1, 2, 3, 5, 8, 13, 21, 34, 55 . . . Fibonacci imagined this using a population of rabbits. You start with one male and one female rabbit. They mate and at the end of the second month the female produces another pair. In this scenario, rabbits never die, but keep reproducing, and each pair produces one new pair – a male and a female who also mate and reproduce. This sequence can grow indefinitely.

If you then look at the ratio between two adjacent terms, dividing one with its predecessor (say, divide 34 by 21), you will get something close to 1.61. If you keep going, dividing ever larger numbers from this sequence with the number that precedes it, this ratio will grow closer and closer to a number, approximately 1.618, symbolised by φ. Fibonacci didn't invent this number; it was already described and known in Euclid's *Elements* in connection to geometry. There it is given as showing that you can divide a line into two parts in such a way that the ratio of the whole line to the longer length is the same as the longer to shorter length.

But Fibonacci showed how the golden ratio connected to numbers. If you take terms from the Fibonacci sequence and create squares with sides of those lengths, you can create a beautiful spiral. This spiral appears in nature in many places, for example in the distribution of seeds in a sunflower's head.

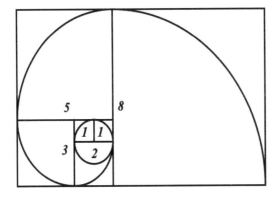

Fibonacci also wrote one of the most important mathematical books of this period, *Liber quadratorum* (*The Book of Squares*), in 1225. It was preserved in a single manuscript and forgotten until it re-emerged some centuries later. It was dedicated to the Holy Roman Emperor, Frederick II, who was himself fond of mathematics. Frederick had grown up in the rich, Arabic-influenced royal court of Sicily (which had been under Muslim rule until the Norman conquest). He discovered that his supposed enemy, Sultan al-Kamil, had the same mathematical interest, and the two exchanged problems in friendly challenges. It is quite remarkable that at this time of deep Christian–Muslim antagonism, the son of the great Frederick Barbarossa and the nephew of Saladin, legendary slayer of European crusaders, should bond over mathematics. It's possible that some of the mathematical problems in *The Book of Squares* came from their correspondence, or were formulated to inspire the emperor.

One of the problems Fibonacci poses is about the product of two sums of two squares. The problem states that if two integers can be written as the sum of two squares, so can their product. For example, we could take 13 (the sum of 2^2 and 3^2: $4 + 9$) and 41 (the sum of 4^2 and 5^2: $16 + 25$). Multiplied together, their product comes to 533, which can also be represented as the sum of two squares, $7^2 + 22^2$ ($49 + 484$) or $2^2 + 23^2$ ($4 + 529$). There's a trick hidden here. If you break down the multiplication of the two sums you could do it in two distinct ways depending on what you first multiply with what and how you arrange their products.

This was not new knowledge – Diophantus proved it some centuries earlier, and Brahmagupta proved a version of it – but Fibonacci and his book made this more widely known in the West, along with much more mathematics he uncovered and revived from ancient Greek and Islamic traditions.

And so, we come to the final part of this period. It was around this time, and around the Mediterranean, that a new mercantile class began to develop. European merchants were distinguishing themselves more and more by the skills they were developing in trade. They were learning from other cultures they encountered and adapting the things they learnt. They could now use Hindu-Arabic numerals and were skilled at quick calculations. This new class needed young men who knew mathematics and knew how to use it in practical applications suited for their trade. This was a sort of mathematical knowledge different from that which was then generally taught, aimed primarily at the upper classes. And so, in the thirteenth century, first in Italy, new abacus schools arose, to which merchants sent their sons. Here the boys would be trained primarily in arithmetic to be used in commerce, and sometimes in some basic algebra, such as solving equations. These developments can't be overestimated in the role they played in the development of trade and accounting. While the new universities flourished and educated the clergy and the ruling classes, the learning of practical mathematics in this period was, particularly in areas of present-day Italy and adjoining French territories, done in abacus schools.

Three distinct branches of mathematics – number theory, algebra and geometry – started to evolve from the old into something quite new. Number theory was revived, the algebra of al-Khwārizmi was learnt and developed, and Euclidean geometry was brought once more to light. Numbers transformed the way that not only mathematicians, but also merchants and everyone who studied them, could use them in calculations. And practical geometry, grasped from Euclid and his followers and translators, started a new cycle where new and old soon collided in unexpected ways. These three branches of mathematics were interwoven in such a way as to enable the birth of early modern mathematics. But not before a couple more strands were added to the mathematical tapestry in the thirteenth and fourteenth centuries.

CHAPTER 11

The Things in Between

Mathematics now entered an era where the road ahead split into two. Huge building works were taking place across Europe, constructing new cities, magnificent cathedrals and ever-expanding universities. All these projects needed sound practical mathematics to underpin both designs and engineering feats of often stunning proportions. At the same time, philosophers and theologians began thinking of the particular importance of mathematics in understanding both this and the other world. Two types of mathematics started to be developed: the practical and the metaphysical.

We don't actually know much about the practical mathematics of the time, although the results of it were becoming immediately and enduringly obvious in the magnificent edifices rising up all around. And there is a very good reason for that. Such mathematics underpinned what were becoming profitable ventures, supporting new building and merchant classes, and so became something of a trade secret. Entry to these trades was very limited, and once you had become a member of a guild where such mathematics was employed, you had a very good chance of achieving success and a prestigious social position – if you kept your knowledge and skills from your competitors.

The other sort of mathematics concerned itself with more ethereal matters, such as space and time. Thomas Aquinas (1225–74), better known for his theological work, introduced an interesting metaphysical thought experiment with a mathematical twist. He was interested in ideas of quantity and magnitude, and posed a question about how many angels could be in the same place. Despite being without a physical body, he said, angels could not occupy the same space at the same time. Later, others made fun of it, recasting it as a question about how many angels could dance on the head of a pin. But it was considered as the prime example of the scholastic philosophy developed in this period.

However, there was a whole body of new mathematics developed now that was neither wholly practical nor wholly metaphysical, but inhabited a space in between, overlapping in ways that are perhaps difficult to understand from our vantage point. One example was the mathematical thinking of Levi ben Gerson, better known as Gersonides (1288–1344). He was a French Jewish philosopher, astrologer, astronomer, rabbi and mathematician, and lived all his life in Orange and the surrounding area. Although we know very little about him personally, we know that he read Arabic and Latin and wrote only in Hebrew. In his wide-ranging work he brought reason, accurate empirical observation and a consummate knowledge of mathematics to bear on theological, cosmological and more ethereal matters. Like Aquinas, he was very interested in the matter of truth and saw that logical reasoning and mathematics have an important place in education.

One of Gersonides' main contributions was improving the techniques of mathematical proofs. He was among the first to use *proof by induction*. We saw in Chapter 9 the inductive reasoning employed by al-Karaji in his arguments. Gersonides went a little further than that. Whereas with an inductive argument you assume a premise to be true and go on to make conclusions, in an inductive proof you need to *prove* the simple cases on which you base your further conclusions. This is a mathematical technique that was given a name only in the nineteenth century, but which Gersonides applied in the fourteenth. It consists of proving a *base case* – the first case of something being true – then applying that to the next case, which is called an *induction step*. Finally, these two steps form a case for proving your hypothesis for *every* case. A *hypothesis* is a proposition of something that we have to check or prove.

Let's have a look how Gersonides did that. There was an ancient question concerning *permutations* and *combinations* that revolved around the letters of the Hebrew alphabet. Combinations and permutations are both ways of describing how we can bring together elements that make up a set; in the first, the order doesn't matter, and in the second, it does. Using the three digits 1, 2 and 3 to create all possible three-digit numbers (without repetition), we would say that we have one combination but six permutations: 123, 132, 213, 231, 312 and 321. (Our combination locks should really be called 'permutation locks'!) This old question was of particular interest to Jewish scholars as they considered that God created only the things he could name: knowing all possible words would enable you to know all the things God could have created. The tenth-century rabbi Donnolo had already listed some of the permutations from Hebrew letters to create all possible words. But Gersonides wasn't very happy with that approach. He wanted to give a formula for all possible combinations and permutations in general terms rather than a list.

Working out permutations gets more complicated when you have a set of *n* elements from which you choose a smaller set of *k* elements, as was the case with the problem Gersonides was working on – you don't make a word using all the letters from an alphabet. The Hebrew alphabet has twenty-two letters, but the words are shorter than that, and of different lengths. Gersonides came up with a number of formulae in his book *Sefer Maaseh Hoshev* (*The Work of a Counter* or *The Art of Calculation*) written in 1322. Using inductive proof, he first stated it by showing how smaller numbers work, and then proved the hypothesis for any number.

First let's look at what would be the number of permutations of a given number of letters – let's say we are looking at five-letter words. We don't want to simply write out all the possibilities and count them. Gersonides used a mathematical way of working this out using the *factorial* function, which is symbolised using the exclamation sign (!). Yes, factorials are exciting! The factorial of a positive integer is the product of that number and all the numbers smaller than it, until you get to 1. So the factorial of 5 is $5 \times 4 \times 3 \times 2 \times 1 = 120$. Gersonides showed that the number of permutations of a set with n elements is the factorial of that number (so $n!$). He also derived formulae for the number of combinations and permutations of k elements from a set of n. For that we will take the k from n and use that difference to divide the permutation of n elements. So we could say that, simplifying things a bit (not counting letters with accents), from the twenty-two-letter Hebrew alphabet we could get $\frac{22!}{(22-5)!} = 3,160,080$ five-letter words.

Just after Gersonides finished this book, another French mathematician was born who would become a great scholar. Nicole Oresme (1323–82) worked on questions of time and space and invented a type of *coordinate geometry* before Descartes, who is usually credited with its discovery (Chapter 17). Coordinate geometry is one where values are plotted on a graph to determine their position.

Oresme invented this early version of coordinate geometry by drawing on the equivalence of tabular values (values presented in a table) and their positions on a graph in his work *De configurationibus qualitatum et motuum* (*The Configuration of Quantities and Motions*, c.1370). He proposed the use of such graphs as a way of plotting values which depend on each other in some way. He then showed how rectangular coordinates, latitude and longitude, and the geometrical figures that resulted on the graphs, can be successfully used to describe and study motion. He showed the change of *velocity* (speed and direction) in relation to time in a graph. This was the first graphical proof of a theorem that is now called the *Merton theorem* (after the Oxford college where mathematicians first discovered it in the 1330s). Oresme showed that a body moving with constant *acceleration* (change of velocity in relation to time) travels the same distance as if it moved at a constant velocity equal to its velocity at the midpoint of a given time period.

Having studied at the University of Paris, Oresme became friendly with the dauphin Charles and held a number of prominent positions. He became a Grand Master of the College of Navarre in Paris and later became canon of the Cathedral of Rouen. When Charles became King of France in 1364 Oresme became his chaplain and councillor. Oresme lived mainly in Paris and translated from Latin into French. He had friends in Oxford too.

If Paris and, more widely, France of the time was a centre of scholasticism in which mathematics was beginning to carve out a new role and new image, Oxford was its second outpost. Normans lived in many areas of Europe, even as far as Antioch, which is today's Syria. In Oxford, in particular at Merton College, a group of monks were working on measuring *qualities* of things – heat, colour, density and light, to name a few of the most important that they were interested in. Eventually they became known as the Merton calculators.

Out of this group arose an interest in the study of *number sequences* and *series*. A number sequence is a list of numbers given in an order of some kind. If you sum all the terms of a number sequence up, you get a number series. And a series can, even if it is infinite, have a finite quantity as a final sum. In that case we can, as we can't use all the infinite terms, say that the number sequence *tends* towards a value. And although we are dealing with infinite series, we can have a sum of a finite value. This is a bit confusing so let's have a look at an example.

The greatest of the Merton calculators, Richard Swineshead (?–c.1340/54), in his *Liber calculationum* (*Book of Calculation*) used a purely verbal proof to show that the infinite sum of natural numbers divided by corresponding powers of two (i.e. 2, 4, 8, 16, and so on) converges to 2. This means that although the series goes to infinity, its sum is 2.

$$\frac{1}{2}+\frac{2}{4}+\frac{3}{8}+...=2$$

Swineshead's verbal proof was very difficult to understand and follow. And he couldn't really help it as mathematical notation was not yet formulated very well at this time. So, Oresme came to the rescue. His genius was in realising that not having notation was one of the problems, so he resorted to using geometrical proofs.

For the same infinite sum, Oresme used first the square of the length of one unit. He goes further to divide areas of such a square in half and creates successive areas of $\frac{1}{2}, \frac{1}{8}, \frac{1}{16}$ and so on. He then adds each of these areas onto the original shape on top of it, to show how these stack up. The argument is that, because we know the area of the original square is 1, and all of the other stacked areas originally stack into one square also, our series sums to twice the area of the original square, or 2.

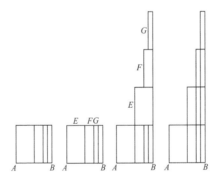

Oresme was one of the originators of the *clockwork universe theory*. This is the idea that the universe is a perfect, precise machine, its mechanical parts all aligned and predictable, which has been set in motion and which will continue working perfectly until the end of time. The obvious question is: who constructed the clock and set it going? This doesn't have much to do with mathematics, but it is important all the same, as it shows that mathematics was often used in the past to describe nature itself. This is a useful perspective, as it offers us a mathematical view of the world around us.

The old mathematics was slowly but surely losing its spell as new ideas and new types of proofs began to emerge. Europe and the Middle East had long been in its thrall, but now new results, and even more importantly, new interests in things like combinatorics, number sequences and series started taking hold. New schools, such as those of Oxford and Paris, and new networks of friends interested in similar areas in mathematics begin to emerge. Between these centres of learning and disparate traditions, mathematics was used to measure qualities and describe quantities. This marked the beginning of the story of early modern mathematics.

CHAPTER 12

Beauty is in the Eye of the Mathematician

Have you ever looked at a Renaissance painting and felt like you were looking through a window directly into the fifteenth century? We have mathematics to thank for it. Since the decline of the ancient Greeks, scenes in paintings had appeared improbable, with flattened figures and unnatural sizes and positions of objects. This all changed in the fourteenth and fifteenth centuries with the development of *linear perspective*, a mathematical technique that renders the illusion of three-dimensional space on a two-dimensional surface. Relatively simple geometry enabled art so wondrous that it enthrals us as much today as it did those who first beheld it.

The system of perspective is based on a few principles. These are relatively simple, but to become very good at it requires months or even years of practice. Imagine you are embarking on painting a Renaissance scene. The basic premise is that there is an underlying geometrical structure to the three-dimensional space we live in (or the space that you are imagining), and we can examine parts of that space. Then there is the frame of your canvas, the vertical and horizontal limits of the space you're looking at, and in that space there are some objects. At this point it can be helpful to imagine you're looking at them from above, and map out where things lie. Back to your easel. Imagine that there is a plate of a kind in front of you, positioned between you and the scene you want to draw. This plate (or plane, or canvas) will be where your image will eventually appear.

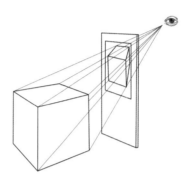

Now the true magic of geometry is applied. From the important points in your scene, you imagine drawing a line that goes from that point to your eye. As it does, it intersects with that plane in between you and the scene. And you can do interesting things with this line connecting your eye and the point of the object. If you wish for your object to appear very far away, for example, you can set the length of that line to be shorter from the eye to the plane than from the plane to the point of the object. It's a matter of ratio. If you wish to be as it were looking down on the scene, you can set the horizon line quite low down on your drawing. The size of the figures will be in some ratio with the size of the objects next to them, and their sizes will be smaller the further away the figures are positioned. The lines connecting objects with the eye would be drawn times over for all the important points in the scene, creating a visual pyramid.

The laws governing perspective construction were first brought to light in this period by Filippo Brunelleschi (1377–1446). The founding father of Renaissance architecture, who created the magnificent dome of the Duomo in Florence, among many other buildings, he used geometric formulae and mathematical modules in his architectural designs. Discovering linear perspective early in his career, he created perspective devices the like of which we have just seen, to help artists of his generation achieve their spectacular realisations of three-dimensional space.

The laws of perspective were set down for the first time by one of the best-known mathematician-artist-architects (and more) from this time, Leon Battista Alberti (1404–72). He was a true Renaissance man (the term could even have been coined for him), excelling himself across a huge range of subjects including linguistics, philosophy and poetry. 'A man can do all things if he will', he is supposed to have said. Alberti's work on perspective is probably the main thing he's known for today, and it was his 1435 book, *De pittura* (*On Painting*), that outlined the principles of the technique that we've just seen (and included a fulsome dedication to Brunelleschi). Alberti made great use of the grid as both an actual and metaphorical aid to artists, which encouraged a new way of looking at the world – geometrically organised in sections. *On Painting* was a revolutionary work not only for artists but for geometers. The investigation of the mathematics underlying the rules of perspective led to the development of modern projective geometry.

The remarkable Alberti, who counted cryptography as yet another string to his bow, also created one of the first polyalphabetic ciphers. This is a cipher where the method of substitution uses more than one alphabet, and for the encipherment, Alberti devised a movable spinning cipher disc where one alphabet could be hidden behind another. The ability to use additional substitution alphabets which could be changed at will was a major advance in the field, and Alberti's system was the strongest method of encryption at this time. Alberti explained all this in his 1467 work *De componendis cifris* (*On Devising Ciphers*), commissioned by the pope's office. Why would you publish on something that relied on being kept secret in order to work most effectively? With the advent of the printing press, more and more mathematicians were publishing their ideas for a living, and ranging more broadly across the branches of applied mathematics.

During this remarkable period, many of the artists who embraced linear perspective were equally fascinated by mathematics.

The Florentine artist Paolo Uccello (1397–1475) was demonstrably skilled in mathematics. Aside from the geometrical construction required in masterpieces of linear perspective such as his 1470 *The Hunt in the Forest*, other examples of his work testify to his mathematical knowledge. He is credited with creating the stunning marble mosaic floor in St Mark's Basilica in Venice, featuring a small stellated dodecahedron. This is a star-shaped solid made up of twelve pentagons that was only properly described 200 years later by Kepler (in 1619) – so Uccello's knowledge of mathematics would have been very advanced. Also suggesting his love of complex polyhedra are his studies of the *mazzocchio*. This was actually a part of fifteenth-century Italian clothing, worn on the head or the neck (both appear in Uccello's paintings), but gave its name also to a polyhedral torus which became known as a test in perspective drawing. The object itself, carefully drawn by Uccello, is a feat of construction in the geometrical sense.

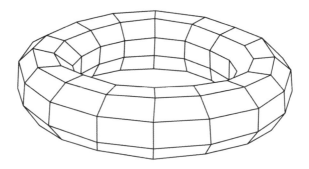

Another such mathematician-artist from this period in Italy was Piero della Francesca (1415–92), today known mostly for his paintings, in which he too excelled at using linear perspective. But Piero also studied the properties of geometrical objects and wrote several mathematical treatises. Some regular polyhedra even found their way into his paintings, as organisational principles for the space. He was particularly interested in the Platonic and Archimedean solids.

We have already seen that Platonic solids are solids made up of the same size and same kind of regular polygons. These were familiar too to Renaissance audiences. Archimedean solids are also made of regular polygons, but not all of the same kind. As their name suggests they were discovered by Archimedes, but over the centuries the knowledge about them had been lost. Mathematicians after Archimedes had described some of them, but no one had defined their properties fully or showed how they looked. No one had made constructions of them either, which was one of the ways of showing that their properties were fully understood.

There are thirteen Archimedean solids and they have rather scary names. Let's look at three of them. The truncated tetrahedron is a tetrahedron whose vertices have been truncated (cut off). Starting from a normal tetrahedron, with equilateral triangles for faces, you cut off the vertices of these triangles in such a way that the main faces become hexagons and the other regular polygons that make up the solid become triangles. Similarly, the cuboctahedron is made of a cube whose vertices have been truncated at the midpoints of its edges, ending up with a solid with faces made up of squares and equilateral triangles. Finally, the truncated icosahedron. As we've just seen, this too is a regular icosahedron whose vertices have been truncated. But because in the icosahedron there are five equilateral triangles at each vertex, when you cut these vertices, you will get a regular pentagon as your second polygon making up the faces. This new solid has thirty-two panels, comprising twelve pentagons and twenty hexagons. Were you to make this from leather and inflate it a little bit, you would get a football.

In *De quinque corporibus regularibus* (*On the Five Regular Bodies*, c.1480–90), Piero della Francesca gave constructions of some of these solids and information on the areas and volumes relative to the measurements of the spheres that surround them. He lent some of his drawings of the Archimedean solids to a young man from his Tuscan birthplace who was also interested in mathematics. He was Luca Pacioli (c.1447–1517).

Pacioli had attended an abacus school where children were taught in the vernacular (not in Latin), and there he learnt the sort of mathematics that would be useful to merchants. His mathematics always had a practical bent. He wrote early books on accounting and bookkeeping, and was the first to introduce the double-entry system, where you record every transaction you make in two lists – one for income (credit) and one for outgoings (debit) – that we still use today when we 'balance the books'. He also wrote one of the first successful mathematics textbooks, his *Summa de arithmetica, geometria, proportioni et proportionalita* (*Summary of Arithmetic, Geometry, Proportions and Proportionality*, 1494), a comprehensive overview of Renaissance mathematics at the time, meant to be used to teach and study. In this work he used items from Fibonacci's *Liber abbaci* as well as al-Khwārizmi's arithmetic and algebraic writings. Did mathematicians give credit to those who invented the mathematics they used in their books, or the discoveries they drew from? This was not always the case, at this time.

Long after his death, an accusation was levelled against Pacioli that he had plagiarised his friend, Piero della Francesca. The third part of Pacioli's *Divina proportione* (*The Divine Proportion*, 1509) turned out to be an Italian translation of Piero's earlier work, the original manuscript of which was found in the Vatican Library. The older mathematician's writings weren't widely known in his lifetime. It's possible he passed his papers to Pacioli as he himself didn't value them greatly, whereas Pacioli was very keen to learn mathematics; or that he wanted his friend to use the things he gave him and publish them. One thing is certain, Pacioli himself had considerable mathematical knowledge, and this book of geometry was a masterwork. Pacioli worked on it with none other than the great Leonardo da Vinci (1452–1519), and it is in this book that the phrase *divine proportion* was first conjured. It refers to the golden ratio.

In an era characterised by the resurgence of neo-Platonism, a philosophy underpinned by the learning and ethics of antiquity, a great many things that were brand new emerged. Mathematics became central to the philosophy of the Renaissance, a word that itself meant rejuvenation and awakening. If the revival of old mathematics provided the spark, the extraordinary thinkers of the period wafted it into a flame.

CHAPTER 13

The Cubic Affair

Just after the turn of the sixteenth century, Luca Pacioli was visited by a mathematician from Bologna, Scipione Ferro (1465–1526), a lecturer in arithmetic and geometry. He wanted to discuss a way he had found to solve *cubic equations* (those where the highest power of the unknown quantity is 3). People had known how to solve linear and quadratic equations for many centuries, but no one could figure cubics out. Pacioli had actually said in the conclusion to his *Summa* that finding a general formula or algorithm for cubic equations was impossible.

The two men discussed what Ferro had discovered. He had somehow come up with a solution for cubics of the form $x^3 + cx = d$, where both c and d are positive. This wasn't a full-scale cubic (which is $ax^3 + bx^2 + cx = d$), but it was still a huge advance on what people then knew. Despite his extraordinary achievement, for some reason Ferro didn't want his formula to be known until he took his last breath. On his deathbed, he duly divulged the secret to his student, Antonio Maria del Fiore.

At about the same time, another Italian mathematician also became interested in the cubic problem. His name was Niccolò Fontana, known somewhat unkindly as Tartaglia, or 'stammerer' (1499/1500–57). As a child he had been injured in the mouth and jaw when French forces had attacked his birthplace of Brescia, which affected his speech, and he had subsequently adopted the nickname as his own. His family was poor and he had a tumultuous childhood, his father being killed when Tartaglia was six years old. As a result he was mainly self-taught, but he rose to become a mathematician, teacher, engineer, surveyor, designer of fortifications and inventor of the science of ballistics.

A few years after Ferro died, Tartaglia was alerted to the problem of solving cubics by a letter from Zuanne de Tonini da Coi, a bookseller from his hometown and himself an amateur mathematician. In that letter Tonini presented Tartaglia with two algebraic cubic equations. And so began Tartaglia's quest to solve them.

In the meantime he had to provide for his family. The year 1534 saw him move to Venice, a centre of publishing and learning and one of the most attractive cities to live in at the time. He was a very good mathematician, but relatively unknown, so decided to start participating in public debates on mathematics in order to boost his reputation and social standing, and maybe win himself a university position. Such mathematical duels were great publicity, and sure enough there was a fight waiting for him.

Antonio Maria del Fiore, armed with the knowledge Ferro had bequeathed to him, made himself known to Tartaglia. The two men began corresponding about their common cubic interest and, in 1535, Fiore challenged Tartaglia to a mathematical duel to determine who was the better mathematician and who could solve the unsolvable problem. They set each other thirty problems and agreed a deadline of fifty days hence to submit answers to each other. Fiore might have been confident at the outset, given that he had Ferro's secret, but Tartaglia had been working on cubic equations for years and the problems he set were of a form his rival did not recognise. Ferro's knowledge was useless. Tartaglia, meanwhile, laboured over the problems before him. Apparently hearing that

Fiore might have a secret weapon, he sat down with a new determination and solved all the problems just days before the deadline – these sorts of miracle breakthroughs often happen to mathematicians! Tartaglia won the contest and became famous overnight, while Fiore disappeared from public view.

News of Tartaglia's victory spread throughout Italy, to Milan, where a physician and mathematician by the name of Girolamo Cardano (1501–76) learnt of it too. He came from a wealthy, privileged background. His lawyer father was famed as being so good at geometry that he was asked to explain it to none other than Leonardo da Vinci, and he taught mathematics to his son. Cardano studied medicine, but took to gambling after his father's death and squandered his large inheritance. Eventually he gained his doctorate in medicine but was refused entry to the College of Physicians due to his bad behaviour. By this point he had become known as a difficult and sometimes violent man, on one occasion even slashing the face of an opponent in a card game. Cardano set up a small medical practice, but wasn't very happy. Eventually, on the advice of his father's friends, he moved to Milan where a lucky break awaited him. His father had held the position of lecturer of mathematics at the Piatti Foundation, and this was now offered to Cardano. It would bring him not only a nice income but a respectable position in high society.

But Cardano was ambitious too. He had heard of Tartaglia and his great mathematical victory and wanted to find out more. In 1539 he invited Tartaglia to visit him in Milan. Tartaglia must have been flattered – a wealthy and well-respected mathematician was interested in his work and had invited him into his home. So, against his better judgement, Tartaglia agreed to show Cardano his method of solving cubic equations, which he had jealously guarded up until then. He made Cardano swear an oath that he would not divulge what he learnt from him to anyone.

Now Cardano was captivated by the problem, and continued working on it after Tartaglia left. There were some difficulties he kept coming across though. When solving quadratics, sometimes you end up with a square root of a negative number. The same thing can happen with cubic equations. Until this point, when mathematicians encountered this, they simply left it aside – it was disregarded as a quirk they couldn't do much about. If you consider equations as representing values from the real world, then finding the square root of a negative number was simply not possible – it would be the same as trying to find the length of a square's side if that square's area was negative. Inconceivable! So as there is no negative area in the world of our experience, why would you trouble yourself thinking about this? Mathematicians left it aside for centuries.

But Cardano didn't think like that. He didn't connect the procedure of finding a square root with real areas. So he wrote to Tartaglia about it. Lo and behold, Tartaglia's attitude had shifted. He was annoyed with himself for divulging the secret of cubics, and considered questions about negative roots to be completely irrelevant. Tartaglia consequently responded to Cardano's letters in quite a rude way, telling him bluntly that he thought what he was asking was completely false and impossible. But Cardano wasn't a man to mess with. He couldn't take rudeness without feeling personally attacked, and the unfriendly feelings now grew on both sides. Tartaglia's insistence perhaps made him even more certain that the case of negative roots *was* a valid problem and one that had to be solved. Cardano realised this was a new category of magnitudes, though he never gave a name to it. This was, in fact, the birth of what came to be called *imaginary numbers* – those of the form $\sqrt{-1}$.

A year later, Cardano employed a servant, a young man by the name of Lodovico (Luca) Ferrari (1522–65). The master soon discovered that Ferrari was excellent at mathematics, and in 1541 asked him to take his place teaching mathematics at the Piatti Foundation. Cardano and Ferrari continued working together on finding solutions to all forms of cubics. Ferrari discovered in the midst of this work how to solve equations with x^4 terms, called *quartics*. But despite all the progress they were making, Cardano and Ferrari could not publish anything because Cardano was under oath not to divulge the secret of the cubics! And cubics came before the quartics, so Cardano and Ferrari were rather stuck with their discoveries.

Sometime in the next couple of years, however, Cardano learnt via Ferrari that Antonio Fiore had already found out something about how to solve cubics from Ferro, and had never published a work on this. Cardano, assured that Tartaglia wasn't the only solver of this problem, now felt free to publish as he thought. He set out to write a book, which was published as *Ars Magna* (*The Great Art of Algebraic Rules in One Book*) in 1545.

What a turn of fortunes this was! The first published book on solving cubics was not written by the man who had so famously proved he could do it in the mathematical duel ten years earlier. Tartaglia felt betrayed. He was furious. He accused Cardano of violating the sworn secrecy of the privileged knowledge he had imparted to him. Ferrari now stepped into the fray to defend Cardano. Tartaglia and Ferrari got involved in an all-too-public dispute, writing some pretty horrendous statements about each other and publishing them for all to read.

There was only one way to settle the argument. A contest was finally called in August 1548 where the two mathematicians could debate their mathematics publicly. It was a huge spectacle, held in the gardens of the Frati Zoccolanti monastery, celebrities descending to see who knew how to solve cubics better, with no less than the governor of Milan as the judge. Tartaglia, an experienced and famous mathematician who had already won his stripes in a similar battle years earlier, was full of confidence. Ferrari was an unknown, lowly former servant. But already by the second day it was obvious that things weren't going well for Tartaglia. He could clearly see that Ferrari knew more than he had assumed. Tartaglia slipped away from the contest, and the victory went to Ferrari. He now became famous and a teacher to the emperor's son, no less. Tartaglia withdrew into obscurity.

Some questions remain unanswered. Why did Tartaglia not publish what he knew in the first place? He did write on cubics in his book simply entitled *Questiti* (*Questions*), but this was only published a year after Cardano's *Ars magna*. We can only speculate why he didn't share his workings earlier. Perhaps he felt that retaining his knowledge would stand him in better stead amidst the extremely competitive fifteenth-century mathematics community, where solutions could win you money and career opportunities. Whatever the case, upon losing the contest with Ferrari, Tartaglia lost his teaching work and retreated from view, writing only towards the end of his life a textbook that could be used in teaching at abacus schools.

And what of Cardano? He wrote some 200 publications during his lifetime on different subjects. Two of these were encyclopaedias of natural sciences; another was *Liber de Ludo Aleae* (1663), the first book on the various mathematical principles of working out likelihoods in games of chance.

Interesting also is his work on special curves called *hypocycloids*. These are curves that are generated when a smaller circle spins within a larger circle. If you took one point on the circumference of the smaller circle and put your pen on it, then spun that circle around the inside of the larger circle, you would generate a hypocycloid. When the point hits the larger circle, it is called a *cusp*. The number of cusps is related to the ratio between the smaller and larger circles. For example, if you take a small circle of a radius 1, and a larger circle of a radius of 4, you will get four cusps. Cardano wrote about this in his book *Opus novum de proportionibus* (*Mathematical Treatise on Mechanics*) in 1570.

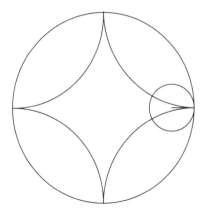

After all that drama, Tartaglia and Cardano both had a rather sorrowful time in their later years. Cardano eventually settled in Rome, but his children were even more troublesome than he had been in his youth – one was executed for killing his wife. Tartaglia, having lost not only his teaching post following his defeat in the duel with Ferrari, but also his family during the plague, died reportedly in poverty in Venice, in his villa near the Rialto Bridge (rather grander circumstances than any ordinary mathematician would find themselves in today).

One of the most sensational chapters in mathematics' long history was over, and as a consequence we have solutions to cubics, quartics, and an inkling that you can, after all, deal with square roots of negative numbers.

CHAPTER 14

Cracking Algebra's Codes

'There is no problem that cannot be solved!' exclaimed François Viète (1540–1603) in his book *Introduction en l'art analytic* (*Introduction to Analytical Art*), published in 1591. You may find this slightly optimistic – but he didn't mean it literally. Viète was the inventor of a system by which he analysed mathematical questions and formulated them in a new way. This new way of doing mathematics he called 'analytical art'.

There is always some question we want to solve with mathematics: how much of something, how soon will something take place, what is the most efficient way to do things, what other shape would have the same area, and so on. Take such a question, and usually you will find that you can formulate it by using algebra. This was what Viète developed to the heights of his new art, and it was groundbreaking when he came up with it. So, when he said that there was no problem that can't be solved, he was trying to say that if you were able to write the problem down using his method, you would be able to solve it. If you don't have a formulation of the problem in the first place, you are much further away from seeing what it is about, and from its solution.

Until Viète, an algebraic statement (for example, find two quantities such that they add up to ten and when multiplied give twenty-four) would be stated just like that, using words. Sometimes diagrams would be added to make things easier to see, or a few numbers and letters here and there. But the mathematical problems would be stated as sentences. Viète came up with a new thing: he would instead write quantities represented by letters, and use the signs such as + (addition), – (subtraction), and a sign for a fraction when division was needed. His statement would be written using these signs and letters to substitute words for quantities and operations.

Viète was a French mathematician and is sometimes called the father of modern algebra. This is not presumptuous; there were algebraists before him, but Viète transformed algebra into the *symbolic* algebra that we recognise and use today. Symbols became a crucial part of what he called the 'art' of constructing and then solving equations. It took some time to gather together all the symbols that mathematicians needed, but Viète was the first to set it all in motion. Since Viète, the balancing of equations and other algebraic procedures could be done without clumsy explanations, but with simple symbols.

Algebra was now becoming the new universal language of mathematics. There was plenty more work to be done to create all the notation needed for this language to become truly functional. And in general, in mathematics, this is the work that still goes on. Coming up with new symbols as you need them for new discoveries remains part of mathematicians' work even today.

Let's just for a moment look at one such symbol. Viète's analytical art dealt with known and unknown quantities arranged in such a way that they were balanced around the equality sign. The = is believed to have been first used by the Welshman Robert Recorde (1510–58). He was appalled how little mathematics was known by the general population in Wales and England at the time. So he started writing his first book with this purpose in mind – whoever would benefit from mathematics would learn how and why from *The Grounde of Artes* (1543). He even listed, in the preface, the particular professions that may benefit from the mathematical knowledge therein. It spurred him to start writing further books on mathematics. As a series, these were *the* first mathematics textbooks written for ordinary folk in the English language – those people who couldn't read Latin or Ancient Greek, or who were disbarred from studying in schools and universities because of their social class. *The Grounde of Artes* was a basic introduction to arithmetic, whole numbers, fractions and geometry and, as with many mathematics books, drawing on ancient Greek dialogues, it was written also as a dialogue between master and a pupil. By all accounts it was a hit, being reprinted in more than forty-five editions all the way to the beginning of the eighteenth century. From this book grew his second, on algebra, *The Whetstone of Witte* (1557). Its aim is there in the title: Recorde was certain that by teaching algebra he would sharpen the minds of his readers. Here he described how tedious he found the repetition of the words 'equal to'. So he invented a pair of parallel lines, one above the other, to replace these two words – because, as he said, 'noe 2 thynges can be moare equalle'. And there you have it, with a stroke of a pen (or rather, two), the equality sign was created, to be used for ever after.

Viète and Recorde were doing fairly similar things with algebra. They wanted to shorten the statements used and make things much easier to write, as well as to make sure such signs and statements had very precise meanings. But strangely enough, given how central algebra is to what we now call pure mathematics, neither was purely a mathematician. Recorde was a physician and controller of the Bristol mint (where coins were made), which perhaps involved some mathematical calculations. And Viète, a statesman and advisor to the French king, also applied his mathematical knowledge in his crucial role of royal cryptographer.

Viète lived in a time of political turmoil when being able to write and decipher secret messages was exceedingly important. European dynasties were fighting one another not only over influence in

Europe, but increasingly over control of new territories now being explored around the world. Enciphering messages and deciphering those of the enemy was crucial to any state looking to found an empire at such distances. But what does algebra have to do with secret messages? Quite a bit as it turns out. In cryptography you replace letters for different letters or even symbols. In algebra you replace numbers by letters. These two things had different purposes but were not that dissimilar in this most important common feature – substitution. Viète used what he had already developed in algebra to make ciphers and crack the enemy's codes.

A brilliant contemporary and compatriot of Viète, Blaise de Vigenère (1523–96), invented a cipher now named after him. He seems to have had quite a lot of fun with it too, teasing his enemies by first writing half of a message in a cipher that was easy to decrypt, and then writing the second part of the message in his own cipher, which was unbreakable.

A simple substitution cipher, known as the Caesar cipher, had been developed in the Roman period. In this cipher you substitute letters for those a certain number of places along in the same alphabet – for example, if *a* will become *d*, every letter will be moved on three places, so *b* will become *e* and so on. Vigenère added to Caesar's cipher multiple further layers of substitution. Each letter in the plaintext (the original content of the message) would be substituted with a letter the place of which corresponded to a keyword. For example, let's look at the text 'little history of mathematics' and use a keyword, HARD, to encrypt it. The first letter, l, corresponds with the first letter of our keyword, H. H is the eighth letter of the alphabet, so we will move our letter l on eight places to s. The second letter, i, corresponds with A in the keyword, the first letter of the alphabet, so the encrypted letter stays in the first place, as i. And so on. This gives us ciphertext that looks like this: sikwse ylztfuf ow phtyhtakljs. To make it even more difficult, with a longer message we could get rid of the spaces between words or add new ones, or use lower-case or capital letters, so there is no sense of the real word lengths or where a sentence begins or ends. Even the same word repeated in the original plaintext would appear differently in the encrypted message. This would make decryption very difficult. With the Vigenère cipher you could only decipher messages if you knew the keyword used to encode the message in the first place. Even using a very short four-word keyword like we have above means that in our twenty-six-letter English alphabet, there could be 26^4, or 456,976, different ways of encoding a message!

While Viète's friend Vigenère created this wonderful cipher, and others we have already mentioned (such as Alberti and Cardano, and earlier, al-Kindi) were skilled in the art of ciphering, Viète was working on another level. He invented modern cryptanalysis, an entirely new method of deciphering texts. He discovered a key to the Spanish cipher, meaning their secret communications could be read, giving the French a huge advantage over their enemy. Consequently, he was promoted, and was to exclusively decipher enemy codes from there on, becoming something like a cryptographer general. This was – obviously – top-secret work. We know of it only because, on his deathbed, he wanted to leave a manual to enable others to make use of his inventions and knowledge. He passed this on to Henry IV's chief minister, the duke of Sully.

Though the treatise is now lost, we know that in it Viète stated the most important rules of cryptanalysis. It was not about looking for specific words or phrases or hoping for useful mistakes or a flash of inspiration. It was a methodical process which perhaps could only have come from the mind of a mathematician. One of his methods was to look for groups of symbols in the ciphertext. For example, with the knowledge that three consonants rarely occur together in French, Spanish and Italian, if you could find some repeating triads in the ciphertext you would know that each would contain a vowel. Identify what might be vowels and consonants and use, where you could, frequency analysis to decipher the text.

Viète's approach to cryptanalysis was similar to his approach to algebra. He reduced both to general rules and procedures, so that once the rules were applied, success always (he hoped) followed. Trial and error could take you only so far; identifying the rules and applying them would take you much further. Whether solving an equation or solving a cipher, the method of substitution, and the use of letters and symbols to denote 'something else', were central to his system.

Viète's work in cryptography contributed greatly to the development of this applied branch of mathematics, today used perhaps less in statecraft and more to protect our sensitive financial data. But more importantly for our history, and the development of mathematics, he transformed algebra, prompting new inventions in its symbols and vocabulary, and enabling calculations of a complexity he could only have dreamed of.

CHAPTER 15

Harmony of the World

We all have moments when we are struck by the sheer beauty of the world around us and the great cosmos of which we are part. The spirals in a seashell, the sweeping curves in a flowing river, the branching of a mighty oak, the intricacy of spiderwebs and the movement of stars in the heavens, all are underpinned by mathematics – which to me makes them even more beautiful.

The planetary movements and their relation to mathematics were something that particularly interested Johannes Kepler (1571–1630). He studied at the University of Tübingen, where he was at first surprised to learn that he was good at mathematics. He wrote his dissertation in defence of Copernicus's 1543 heliocentric theory, which had established the Sun as being at the centre of the solar system. Many scholars and particularly Church authorities still fiercely resisted this idea. So Kepler couched all this in a curious thought experiment. He imagined that he was able to travel to the Moon. If he could, he said, he would be able to see that the skies there were not too dissimilar to those above Earth.

Looking up from both places you would see the same, or nearly the same, stars and planets and their movements. But on the Moon, you would also see the Earth changing position, just as on Earth we observe lunar motion in the night sky, because he knew that the Earth travelled around the Sun, and the Moon around the Earth, though we can't feel our planet speeding through space with our senses.

To say this today would not be considered strange, but in Germany at the time it was unusual and brave. And although it was almost four centuries away in the future, travelling to the Moon was an idea that stayed with Kepler. Later, in 1608, he expanded it in a little novel, *Somnium* (*Dream*), which imagines a flight to the Moon and the problems you need to tackle, from finding a good moment for take-off to dealing with direct sunlight, extreme cold and lack of air, though he didn't dwell too much on the technology required for space flight. Nevertheless, the idea sparked the start of his work on planetary motions.

In 1596, after finishing his studies, he wrote a book, *Magisterium Cosmographicum* (*Cosmographic Mystery*), where he described what he thought the structure of the universe was really like. The universe, he said, behaved in a mathematical way. Through mathematical knowledge one could learn of its structure and laws. In his book Kepler presented an image of a model of the universe where each of the five Platonic solids was encased in a sphere and then inscribed in another solid. The intervals between the spheres and objects, he suggested, corresponded to the sizes of each of the (then known) planets' orbits around the Sun. This wasn't entirely true, and he soon realised that. The planets didn't move in circles. He realised this was an error and worked and thought further on it. But this work led him to consider the mathematical laws that underline cosmic laws.

Firstly, he worked out that the orbits of the planets around the Sun were elliptical. While a circle has a centre, an ellipse has two *foci* (special points used in geometry to construct curves). The sum of the distances between an ellipse's foci and any point on the ellipse always remains the same. If you attached a piece of loose string to two points, then tightened the middle of that string with a pen and moved the pen around the two foci, you draw an ellipse. The orbits of each of the planets in the solar system are elliptical, with the Sun at one of their foci. This was Kepler's first law of planetary motion, now called Kepler's First Law.

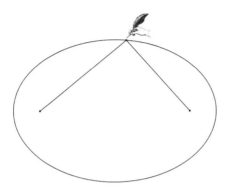

He then looked at how planets move on their elliptical orbits. He noticed that they move faster when they are nearer the Sun. He calculated how a planet's speed changes depending on its position in its orbit. A planet 'sweeps out' the same area of space in the same amount of time no matter where it is in its orbit. This was his second law of planetary motion.

From the start of the 1600s Kepler was living in Prague and working with a famous Danish astronomer, Tycho Brahe. Brahe had recently been appointed Imperial Mathematician to Rudolph II, the Holy Roman Emperor, who was interested in the arts, sciences and astronomy in particular. Rudolph had asked Tycho to compile tables of the movements of all the visible stars. No sooner had Kepler joined Tycho to assist with the mathematics on this grand project then Tycho died. Kepler succeeded him as Imperial Mathematician and remained in Prague for some twelve further years, still diligently working on this task. The ambitious aim was that the tables would locate the possible positions of the Moon and the planets of the solar system for all time – past, present and future.

Kepler didn't only notice things in the heavens above us. He noticed things that fell to Earth too. In 1611 he wrote a rather lovely little book about the shape of snowflakes. He conjectured there (as he couldn't prove it) that hexagons are the best way of packing in a plane. The bees knew this already, but Kepler worked out that this must hold in general. (We will hear more about this in Chapter 39.)

Kepler saw beauty in everything around him, despite huge problems that he faced throughout his life, such as recurrent wars and family misfortunes, including his mother being accused of being a witch. *Musica universalis*, or the music of the spheres, was a philosophical concept that had been discussed since the medieval period that suggested that planets dance to an unheard and unknown piece of divine music. Kepler started a book on precisely this, straight after his *Magisterium Cosmographicum*: he wanted to show that there really were harmonic proportions in the spacing and movements of the planets, and to describe this harmony precisely through mathematical formulae.

The book that came out of this research was *Harmonices Mundi* (*Harmony of the Worlds*) in 1619. In it, Kepler studied the speed at which the Earth goes around the Sun. He realised this speed changes: Earth is faster when it is closer to the Sun, and slower when it is further away from it. He worked out that this varies by the ratio 16:15. This has an equivalent in music, where sounds produced by ratios of the lengths of chords can be seen as special ratios themselves (Chapter 4). This particular ratio is equivalent to a semitone in music, as for example from *mi* to *fa*. Now Kepler imagined a celestial choir made up of the planets: Mars was a tenor, Saturn and Jupiter both bass, Mercury a soprano, and Venus and Earth altos. They would have all sung together in perfect harmony, he said, just once in history, at the time of Creation. As wonderful as this all sounds, it's not entirely accurate, just as his model of the universe wasn't. But Kepler's investigations did reveal that the time it takes for a planet to go around the Sun (its orbital period) is proportional to the size of its orbit: more precisely, the square of the orbital period is proportional to the cube of the semi-major axis of its orbit. This was his third law of planetary motion.

Harmonices Mundi also made important strides in geometry. In the second chapter Kepler described for the first time two new star polyhedra. Although images of these had appeared here and there before him, Kepler was the first to recognise that they are semi-regular solids, additional to the Platonic and Archimedean solids. They formed a new little set of semi-regular polyhedra: the small and great stellated dodecahedron. The pyramids making the objects into a star (stellation) are set on a pentagonal basis. Underneath these stellations are pentagonal faces.

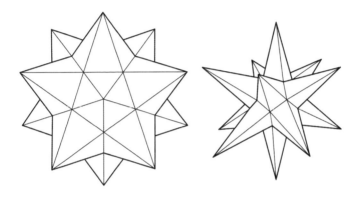

The book was a wonder of mathematical discoveries, although they were all presented as being there to deliver the message of the beauty and harmony of the physical world. And, considering that Kepler was deeply religious, all this beauty was the result of divine actions. Kepler thought that God made the world according to mathematical principles and that to understand God he had to understand mathematics.

All the while when researching and writing *Harmonices Mundi*, Kepler had kept working on that task he had started with Tycho. It had been nearly two decades since Tycho had died, and Kepler was still struggling with those tables and all the things he was trying to do with the measurements of planets and stars that Brahe had accumulated. Part of the problem was how to deal with the arduous calculations of the huge numbers arising with such a vast cosmos. If only there was a computational method that could speed things up.

In 1614, the heavens answered his plea: logarithms were invented. When Kepler first learnt of them in 1616, in John Napier's *Mirifici Logarithmorum Canonis Descriptio* (*Construction of the Wonderful Canon of Logarithms*), he described their invention as a 'happy calamity'. Why his joy was so circumscribed we can but guess. Could it be that now he had no excuse not to complete the tables?

The word *logarithm* comes from two Greek words: *logos*, meaning ratio, and *arithmos*, meaning number. So, the ratio of numbers, but in a very particular way. This is how logarithms work. You can multiply a number by itself many times and you get a product. For example, $2 \times 2 \times 2 \times 2 \times 2 = 32$. Or you could write that as $2^5 = 32$. There are three numbers here: 2, 5 and 32. We started from 2, put it to the power of 5, and got 32. The number 2 we call the *base*, as it was the base number you put to a power. And a logarithm is the power to which a base number must be raised to get a given number. So in this example, we would say that we want to find the logarithm of 32, base 2, and get the answer 5. In mathematical symbols, $\log_2 32 = 5$.

This was a fantastically useful invention as it allowed you to replace multiplication and division with addition and subtraction through a procedure. Let's see how. The logarithmic tables made by Napier also connected two types of numbers – natural (or counting) numbers, and those which are powers of something. Let's stick with the powers of 2 for our purposes. The top line gives indices (here, natural numbers), and the bottom row are the powers of 2:

1	2	3	4	5	6	7	8	9	10
2	4	8	16	32	64	128	256	512	1,024

Now, the bottom row also represents a geometric sequence, while the top represents an arithmetic one. An arithmetic sequence is one where a difference between terms is the same number (added or taken away to the next term). A geometric sequence is one where you multiply (rather than add) each term with a common ratio to get new terms.

Let's now take two numbers from the top row, say 4 and 6. If we add them up we get 10. Now look at the corresponding values under 4, 6 and 10. They are 16, 64 and 1,024. If you multiply 16 and 64, you will get 1,024. That is because there are laws of powers and logarithms, and the one we just applied tells us that, as $a^4 = a \times a \times a \times a$, then $a^4 \times a^6 = a^{(4+6)} = a^{10}$. Having tables such as those Napier calculated and wrote enabled people to look up the logarithm of *any* number. To multiply two numbers they could then look up the logarithms of those numbers, add them together, and then use the tables in reverse to find the final answer.

Imagine struggling for years to make some sense out of calculating with huge numbers, and then suddenly having access to such tables. Kepler liked them, but not entirely. He already knew another mathematician, Jost Bürgi (1552–1632), a Swiss clockmaker and maker of mathematical and astronomical instruments, who came to work at the court in Prague. He had invented logarithms some years earlier, and independently of Napier, though he hadn't yet published on them. They seemed to have a tricky relationship – Kepler later criticised Bürgi as 'an indolent man, and very uncommunicative', saying that 'instead of rearing up his child for the public benefit' and publishing his logarithmic method, 'he deserted it at birth'. But the two men did discuss logarithms together at Prague. Kepler asked Bürgi to redo the logarithms for his specific purposes, but the results were difficult to understand, so Kepler decided to make his own. In *Ephemerides* (1620) he described how these new logarithms worked; the method was just shorter, the calculations and meaning very much similar. He dedicated this little book to Napier, the main inventor of logarithms, but made his own logarithmic tables more relevant to the planetary motions he was studying.

Kepler eventually finished his (and Brahe's) masterpiece, which he titled the *Rudolphine Tables* in honour of the Holy Roman Emperor who had commissioned them. This was a feat of incredible proportions. Here we find a catalogue of the positions of 1,405 stars, directions for locating the planets of the solar system, and also an incredibly precise map of the world, instructing how to use lunar distance to measure longitude. There were also, of course, the much-anticipated logarithmic tables, computing the precise times the planets could be found as they moved in their orbits around the Sun. The first test of the accuracy of these calculations took place on 7 November 1631, the date that Kepler had predicted Mercury would transit over the disc of the Sun. He was proven right, his tables accurate to a few hours. But by this time, he had been dead for almost a year.

Kepler was the greatest master of calculation of his times. He had a great vision of the beauty of the universe which he uncovered through mathematical laws. An interesting feature of his work was also that he used his error – the one about the shape of the orbits of planets in the solar system – to make new discoveries, refining and improving his first model of the universe until he defined the three laws of planetary motion, which we still use today.

CHAPTER 16

Mathematical Machines

That 'happy calamity' that helped Kepler with his big numbers, the invention of logarithms, was not an accident. Trying to find something that would make calculations easier to deal with became all the rage in this period. Not only mathematicians but merchants, astronomers and explorers all needed some sort of calculating system, and even dreamed of making machines to aid them in their mathematical tasks.

Seventeenth-century mathematicians were all too aware of this. It was they who were most often involved with tedious repetitive calculations where they just needed results, rather than the calculation itself being important. This was, frankly, boring work. Some, it could also be said, were excited by the prospect of dreaming up some new mathematical toys they could play with. Whatever the reason, suddenly there was an interest in devising new things that could help calculations, and mathematicians started imagining mathematical devices and machines that could do the work for them.

John Napier (1550–1617), whom we briefly met in the last chapter, was the first to come up with something. His logarithms arose from another of his inventions, one that dealt with long multiplication – a device he invented in 1614, the same year he published his *Mirifici Logarithmorum*. It was a set of rods that you could do calculations with. The most expensive sets were made of horn or ivory, and hence they became known as Napier's bones. These long rectangular rods were divided into squares. One rod had the numbers from 1 to 9 or 10; each of the others had one of the counting numbers at the top, and multiples of that number down their lengths; the two-digit multiples were written across the diagonal line that divided the squares. Positioning the rods, and reading them in a certain way, lets you do quite complex multiplication of large numbers by adding and carrying over, and requires considerably less thinking than doing the multiplication long-hand. The rods do the multiplication for you.

Napier's logarithms, which he invented after his rods, inspired the English mathematician Henry Briggs (1561–1630) to come up with his own version. Napier's formulation was a bit awkward, so

Briggs simplified things by looking at the logarithms with base 10. They are called *common* or sometimes *Briggsian logarithms*. But Briggs made tables of precalculated logarithms of many different numbers. His *Arithmetica Logarithmica* of 1624 contained the logarithms of an astonishing 30,000 numbers, and his tables were in use for a very long time, all the way into the twentieth century.

Page after page of tables of numbers: if you looked at them without knowing how to use them, they wouldn't mean much. But you could do all sorts of things with them. For example, the tables allowed you to find square roots. We saw in Chapter 15 how logarithms could be used to transform multiplication and division of large numbers into addition and subtraction. Briggs also provided logarithms of small prime numbers (up to 100). We have known since Euclid's times that there are infinitely many prime numbers, so he couldn't provide logarithms of all of them. Still, this was very useful. Why? Well, because each whole number can be expressed as a product of its prime factors. You can break any whole number down in this way. And so if you have logs of these primes, you can simplify whatever calculations you want to do. Today, we have calculators on our computers and even the phones in our back pockets, but before this, having tables with precalculated results, and knowing how to use them, was the fastest and easiest way to work calculations out.

The greatest and possibly most important application logarithms had, however, was in navigation and ballistics (the study of projectiles and their flight paths). All the new emerging professions in the employ of the new empires – scientists, engineers, navigators – made great use of this wonderful little invention to simplify their calculations and achieve greater precision as well as efficiency. Instead of getting bogged down in long and protracted calculations, they could now concentrate on finding ways to improve their crafts. One mathematician once called logarithms one of the greatest scientific discoveries the world had ever seen. On the back of these tables were built empires.

Rods and tables are all well and good, but where were the machines that mathematicians dreamed of? It was Wilhelm

Schickard (1592–1635) who invented the first mechanical calculating machine, in 1623. Its design was concocted from a version of Napier's bones which could be rotated, and a few other things. It had clock wheels and gears. Each digit had an input wheel (for the number you wanted to calculate something with), an intermediary wheel to carry the transfer of the calculation itself, and a display wheel, which all meshed together. Well, that is what the design drafts say. Schickard asked a clockmaker to make the machine for him. Whether they got stuck on it, or it didn't work well, for whatever reason the machine was never finished. Then what was made was destroyed in a fire, and Schickard abandoned the project altogether. Schickard's calculator therefore never saw the light of day. We know of it because some sketches for it were discovered in the twentieth century, and some examples of the working machine were made subsequently – you can find them in science and mathematical museums around the world.

Not too long after that, Blaise Pascal (1623–62) had better luck. He managed to design and construct a real calculating machine. And it worked! By all accounts Pascal was a child prodigy. His father had forbidden him to do any mathematics before the age of fifteen, so of course, Pascal simply had to become a mathematician. Pascal invented his calculator when he was somewhere between nineteen and twenty-one years of age, and the machine was designed to help his father do his work as a tax collector. It was the first machine to do mathematics in a mechanical way. It had nine dials and a carry mechanism which can carry 1 to the next dial (as we do when we add by hand as well). Some of these calculators were made for sale – maybe only a hundred or so, as it wouldn't have been a cheap gadget to buy. Surviving copies, now called the Pascaline, can also be found in museums today.

Why are these devices and methods important for mathematics? Crunching numbers seems to be all there is to it. But this trend to start thinking about how to make shortcuts with mathematical processes is not only to be found in the work with numbers. In this period of global exploration and warfare, there was a heightened need to build ships that would withstand long voyages and outmanoeuvre the enemy, as well as a desire for machinery of more complexity to meet the demands of an increasingly mechanised world. Precision was becoming increasingly important for engineering, astronomy and navigation. In all kinds of ways, mathematical work was increasing in quantity and scope.

A great number of new engineering creations were being developed, with the Dutch leading the way. They exported their engineers to faraway places, from England to Japan, and they in turn needed mathematical instruments. Such instruments were made and improved through knowing the mathematical laws underlying their every function. Geometry, for instance, was a critical skill in engineering. How do you construct a curve that would precisely describe the best curvature of a ship? And could some machines be made to do that too?

The Dutch mathematician Frans van Schooten (1615–60) looked at the mathematical machines that were being made for performing calculations, although he was mainly interested in geometry. So he set himself the task of making machines that, given different parameters, would be able to draw different *conic sections*. Conic sections are curves that we get when we 'cut' a cone by a flat plane. There are several ways of doing this, which give us different curves. (We only really count the parabola, ellipse, circle and hyperbola as curves here, which are called *non-degenerate* conic sections.) Some curves you can't construct mechanically, but conic sections you can.

Van Schooten set out by first studying the structures of conic sections and thinking about the laws that govern them, so that he could use that knowledge to make machines that could draw such curves. Let's look at how he designed a machine to construct a parabola. A *parabola* is a curve resulting from quadratic functions (of the form $y = ax^2 + bx + c$). Every parabola has some parts we need to name – the focus (the point that lies on the parabola's axis of symmetry, labelled B) and the vertex (where the parabola crosses its axis of symmetry, labelled A). There is a line above the parabola (labelled E) that is perpendicular (that is, it makes a right angle) to our parabola's axis of symmetry. This line is called a directrix. In van Schooten's machine, this horizontal line will direct the movement we are about to make. If you put a pin in the directrix (labelled G, at the intersection of two perpendicular lines) and a pen at the point labelled D, and move your hand with the pin horizontally, the pen would draw a parabola.

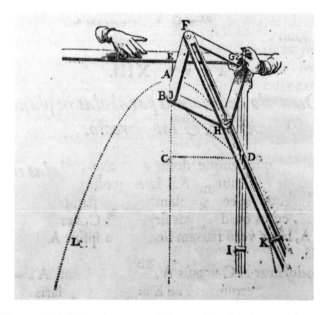

The special thing about parabolas is that the distance between the focus (B) and a point on the line (D) is always the same as that between the pin on the directrix (G) and that point on the line (D).

As with the parabola, all other conic sections can be drawn by continuous motion like this. You need to know how to make the structure that will produce the particular section, but once you work that out you can draw them perfectly. Earlier, we saw how to draw an ellipse using a string, two pins and a stretched string. Van Schooten, as we've seen, was more ambitious, and wrote a book describing the many kinds of devices needed to produce conic sections.

Mathematics in this period was becoming something much more than just a theoretical science. Mathematicians – and the many others who benefited from mathematical calculations and drawings – needed some faster and more precise ways of doing things that were increasingly becoming difficult to do. These sorts of machines for numerating and drawing sped things up and reduced error – it is much more likely you will make a mistake if you have to do the calculations you need all the time, and it takes your focus away from the task at hand.

But there were further consequences to these developments too. Because mathematicians and engineers didn't have to do tedious work anymore, or at least less of it, they had more opportunity to invent new things. More than ever before, they started questioning things they observed. Mathematicians could now also use these new inventions to check things they conjectured on. It was becoming obvious that there was a certain type of mathematical process that could be mechanised. But there were other things that couldn't. Why not? These types of questions – and answers – became ever more significant, as we'll see in Chapter 20.

CHAPTER 17

Before the Internet, There Was Mersenne

Marin Mersenne (1588–1648) was a Minim friar, a Catholic priest of the Minim order. This order, as its name suggests, had something to do with the minimum of things, the smallest of the small. Their way of life was characterised by abstinence and the importance of being humble.

So it's funny that a man who was devoted to the minimum in life came up with a system that gives us the largest prime numbers we know. The largest currently known prime number discovered in October 2024 is $2^{136,279,841} - 1$. It has a staggering 41,024,320 digits. How long would it take you to count to this number? Well, there was an experiment someone did for a charity in 2007, counting out loud to 1 million. It took Jeremy Harper from the US eighty-nine days, spending sixteen hours each day counting. Multiply that by almost twenty-five and it would take you at least six years to count up to the largest currently known prime number (all those enormous numbers would take much longer to say aloud). I don't think anyone would volunteer to test that out, but you never know.

This number is one of what we now call the *Mersenne primes*. Mersenne conjectured that the numbers you can write as $2^p - 1$ are prime if p is a prime number itself. He wrote this in the preface to his book *Cogitata Physica-Mathematica* that he published in 1644. He was partially right – there are primes which fit this formula, but not always. Because we struggle so much in finding any kind of order or pattern with primes – something that, if you are so minded, would make a very good mathematical project: no one has yet been entirely successful! – Mersenne's formula is still good to have. Close enough. It appears that Mersenne primes are a useful vehicle for searching for and discovering the next largest prime we have so far.

Mersenne is probably the most famous of all Minim friars in history. He lived in his cell in the Minimite monastery in Paris for twenty-nine years of his life. Rarely would he venture out. Yet his outreach was astonishing – he corresponded regularly with people across France and Europe – Holland, Italy, England, Spain – and even in Asia Minor. There exists a saying in mathematical circles that before the Internet, there was Mersenne – if you wanted to find out who was doing what in mathematics, you just had to ask him! Everyone was telling him about their latest mathematical discoveries, and his correspondence circle became a kind of school. He saw this himself, coming to call it the Académie Parisiensis, although it became better known as the Académie Mersenne. It was founded in 1635, and at the beginning of its existence, members of this circle were meeting in each other's houses, but as Mersenne's health became fragile, those who lived in Paris started gathering in his little cell in the monastery. The academy only came to an end in 1666 when one of the circle co-founded France's illustrious Académie des sciences.

The organising of this academy was Mersenne's other major achievement. Nearly 200 scholars engaged in correspondence with Mersenne on a wide range of subjects. He became something of a clearing house for discussions about new knowledge at this time. He would get the letter from a correspondent, see what it was all about, and then pass it on to someone who had a similar interest.

Medicine, law and theology were all discussed, but about a quarter of the academicians were working on mathematics, and among them were some exceedingly important mathematicians in their own right. We will look at just some of these, and how their discoveries were not only important but changed mathematics forever.

The pre-eminent academician at that time was certainly René Descartes (1596–1650), a mathematician, philosopher, scientist and sometime mercenary soldier. Descartes was born in a little town in central France, La Haye en Touraine (which is now called Descartes in his honour). He moved around a bit, corresponding with Mersenne while he travelled, and in Holland met important mathematicians and studied how they applied mathematics to engineering. He ended up in Stockholm, teaching philosophy to Queen Christina of Sweden, and died there later. The main mathematical idea Descartes came up with has been crucial for the development of both mathematics and engineering ever since. This was the idea that through a mathematical system we can precisely determine not only the position of every point we can imagine, but also curves and shapes, through their algebraic equations, which could then be plotted using the coordinate system. Shapes could now be precisely determined and communicated to others who might be making things from those shapes. Until then, the process would be to describe or draw a shape, and then use templates to replicate it. But now an equation would be enough! Well, that was to be further developed over the next few centuries, but the mechanism was suddenly there to do precisely that.

Descartes' mathematical work was part and parcel of his philosophy. His famous quote, 'I think, therefore I am', came about from his original system of methodical doubt. Descartes wanted nothing less than to revolutionise the entire scientific method, and set out new rules in his major work, *Discours de la méthode* (*Discourse on the Method*, 1637). These included breaking down problems into their simplest parts, starting from those in order to work towards a solution, and checking reasoning at all stages. It was the mathematical procedure of deductive reasoning writ large: how to think in a scientific way and how to do this through logical steps.

Descartes also provided various explanations to show how to do this. In his little book *Géométrie*, which was originally an appendix to his great *Discours*, he illustrated how to think using mathematics. He wanted a method that would give you absolute certainty when investigating a mathematical object. What he came up with is to link geometrical objects with their algebraic equations.

He did this through a simple device: he invented a coordinate system. We saw in Chapter 11 that Oresme had already come up with that idea for a particular example. But Descartes now came up with a general system that could be used to define any point on a flat plane and align numbers and lines in any case. Here, a number has a position on a number line. Two such lines are set perpendicularly to each other, the horizontal line called x and the vertical y. Given two numbers – one on the x and one on the y number lines – you can plot a point on the plane. Then if you have two points which you know precisely like that, you can draw a straight line through them. And this straight line will relate to an equation which is given with these two quantities that are in some relationship, x and y. Let's see how that works when, for example, $y = x + 2$.

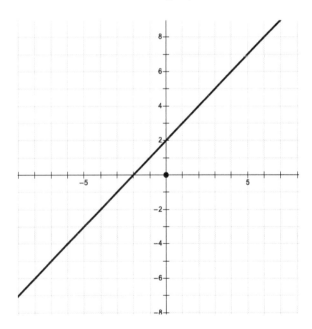

One of the curves he described in 1638 is what is now called the folium of Descartes. This is a lovely curve, folding as it does around the x and y axes. Its equation is $x^3 + y^3 = 3xy$ – quite a simple equation, really, considering the complexity of the curve, and how difficult it would seem if you didn't know how to connect it with its algebraic description. Every pair of values that would satisfy this equation would give you a point on the coordinate system, where x and y coordinates are those values, and would allow you to plot the folium.

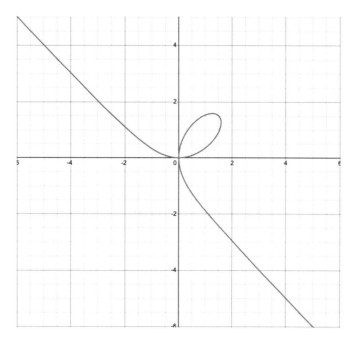

Mathematics played a central role in Descartes' work. In his view it was only through mathematics that investigating things in general could happen precisely. He connected geometry and algebra to create analytic geometry. This system has been crucial to the development of mathematics. It transformed the study of shapes. Suddenly you could not only draw and imagine different shapes but define them through equations.

Another correspondent of Mersenne who was interested in similar things was Girard Desargues (1581–1661). Desargues was both a mathematician and engineer. Like some others before him who were interested in mathematics that could be applied to engineering, he worked more on geometry that didn't necessarily need equations. He had seen how Italian mathematicians of the Renaissance worked on perspective and the mathematics that was contained in it. Now Desargues wanted to invent something that was easier to actually use. You see, the difficulty of using perspective drawing to create an object in the real world is that the lengths and angles may not be easy to work out from the picture, where they are shortened. And this becomes even more difficult with curves. Desargues thought about that. He posed a question about how one could draw a conic section in perspective, but preserve its true lengths and angles. Desargues didn't achieve his aim of coming up with a new entire system, but the way that he played with the question, and actually having asked the question at all, was incredibly important for the development of mathematics.

Desargues drew various conic sections and looked at how conic curves are in some way connected. Imagine a cone and cut it with a flat plane. You may get a circle. Now keep that plane, and then cut the same cone with another plane but at a different angle. These two planes will intersect in some line. What Desargues then did was to reduce this type of thinking about cones and curves to simple lines. At some point he reduced the curves to something simpler too. He looked at triangles. If you have a triangle on one of those flat planes and then another triangle on another flat plane, something strange happens.

Let's say you have two triangles, *ABC* and *A'B'C'*, with corresponding sides (*AC* corresponds with *A'C'*, *A'B'* with *A'B'* and *CB* with *C'B'*). Now if you extend the lines on which those corresponding vertices lie (*A* and *A'*, *B* and *B'*, *C* and *C'*), they will intersect at a single point. And if you extend the segments of triangles (until *AB* meets *A'B'*, and so on), these intersections will all lie on one line (called the axis of perspectivity). This became known as Desargues' theorem.

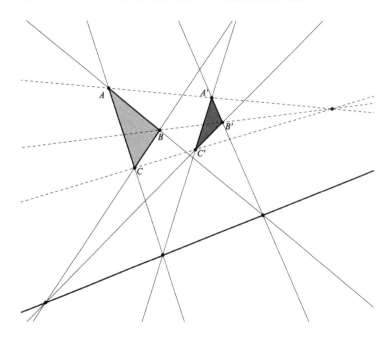

It is amazing because it is quite difficult to get three points to align. A line is determined by two points. If three points are aligned, it is a sign that something important is going on. Desargues didn't go much further with this, but it opened the door to an entirely different type of geometry, now called *projective geometry*. Projective geometry is a type of geometry where the dimensions are not really of much importance. It studies the properties of figures which stay the same as they are *projected* from one plane to another. You could say the two triangles above are like this – one could be original and the other its projection.

Mersenne's academy was a hotbed of mathematical discoveries. By forging connections between mathematicians, enabling them to communicate, and making this a regular occurrence where people could discuss new ideas, Mersenne sowed the seeds of remarkable mathematical inventions and discoveries. Within such a dynamic environment, people from different backgrounds and even countries could exchange information not only about their interests but what they came across in their own circles. What mathematicians had learnt and were wondering about they could put to the test, and pose each other questions that would lead them to areas they may not have thought of themselves otherwise. What this brought to mathematics was a kind of joint effort to test new ideas. Through this channel, members could share discoveries, collaborate and provide feedback to each other, be that encouraging or, as was sometimes the case, a judicious pointing out of mistakes!

This idea that mathematicians work best as part of an international network only grew after Mersenne, never really to be completely lost. But we have not yet mentioned something extraordinary that happened between two members of Mersenne's academy, which brought yet another whole new area of mathematics into existence.

CHAPTER 18

The Game That Never Stopped

Imagine you're playing a game of chance with another person, such as rolling dice and seeing who gets the highest number. You've each put the same amount of money into a pot and have agreed that whichever of you wins a certain number of rounds takes the lot. But something happens to interrupt your game and you have to stop playing before either of you has won. How would you decide how to apportion the money? If you were on the cusp of winning, you wouldn't be too happy to just call it a day and get your money back. That wouldn't feel fair. But how could you know how the game would have gone, and bearing in mind who has won the rounds so far, how can you divide the pot fairly?

This problem had puzzled and intrigued mathematicians for some time. Back in 1494, Luca Pacioli (Chapter 12) had described it in his book *Summa*, and there called it the problem of the unfinished game. His solution was simply to divide the pot in proportion to the number of rounds each player had won. But what happens if the game stops after a single round? The winner of that one round would take all the money! Niccolò Tartaglia (Chapter 13) noticed that rather unsatisfactory and counterintuitive outcome a few decades later. He came up with a new suggestion of dividing the pot according to the ratio between a player's lead and the total number of rounds that should have been played. This was better, but still not right. It didn't take into account whether the player in the lead was close to winning (in a game where the winner is the first to 100 rounds, there is the same ten-point lead with scores of 58–48 as well as 98–88, but in the former the game could still go either way, whereas in the latter a win seems pretty certain). Tartaglia himself wasn't happy with his solution. At this time, games of chance were thought to be essentially incalculable. Was it even possible to work out a system of dividing the pot when a game of several rounds must be abandoned?

On 24 August 1654, Blaise Pascal (who we briefly met in Chapter 16 inventing his calculating machine) wrote a letter to his countryman, the mathematician Pierre de Fermat (c.1607–65), on precisely this topic. It had apparently been suggested to him as a subject worthy of study by the Chevalier de Méré, an inveterate (and highly successful) gambler. Pascal sought Fermat's ideas on solving the problem, and suggested that they should correspond on it. He set the rules of their own game: they were to look at the problem from every angle and pull no punches, feeling free to criticise each other's reasoning when there was a need for it. Pascal himself, one of the world's greatest mathematicians, already had some ideas, but was full of doubt.

So their mathematical play commenced. At first, they too thought that the pot should be divided according to the ratio of points the players had accumulated when the game was interrupted. But eventually it occurred to them that this cannot be correct: the game could have concluded in many different ways, and the division should take into account those specific possible cases. Their great insight was to look not at the history of the game that had already taken place – the number of points each player has from the rounds played so far – but rather the possible ways the game might have progressed in the future, and the probability of each player winning depending on the rounds still required to win.

They put forward two possibilities. One was that the players should agree to play three more rounds. In that case it would be fairly straightforward to work out all the probabilities of the various outcomes and divide the money appropriately. But Fermat said that this is not what happens in life – players tend to want to play until one of them wins and no more. That is a little more difficult to work out.

So Fermat and Pascal set their sights on understanding and enumerating the ways that the game might have continued had it not been stopped. They tried various ways of mathematically describing this. You could do a lot of things, it turned out, and the two men discussed all their ideas at length. They settled on providing a solution based on how many games each player should have had to win in order to have won overall. So you could, for instance, simply list all the combinations of all possible ways that the game could have played out. Fermat proposed something like this in a fully tabulated solution, which was right, but was rather complicated, and included probabilities for rounds that would never have happened (because one of the players would already have won). And how would it work if there were many, many more rounds to play? Pascal and Fermat wanted to get to a set of numbers representing the likelihood of either player winning based on whatever number of games it would take for them to win. From there you would be able to ascribe a numerical value to their chances, make a ratio of their chances, and divide the spoils in that ratio.

This is where Pascal's triangle comes in. Pascal wanted to simplify his friend's unwieldy approach, and what better way to reveal hidden numerical patterns than to set out your numbers in a pattern-like way? Pascal's triangle bears his name today, but others before him, such as eleventh-century Chinese mathematicians, had used something similar to it (though not to explore the question of probability). But now Pascal collected what was known about the triangle and applied it in a completely new way to the problem of probability. This he described in *Traité du triangle arithmétique* (*Treatise on the Arithmetical Triangle*) in 1665. This triangle starts from the top with 1, and the number 1 appears in all the cells along its left- and right-hand sides. Under the first 1, on the first row proper, there are two more 1s. The second and all subsequent rows are made up by adding the two numbers above a cell to get its number. Pascal's triangle can be constructed for numbers as large as you like.

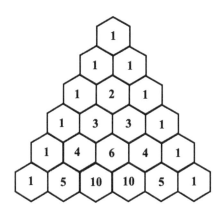

How does it relate to probability and all the new things Pascal was trying to do with it? First let's remind ourselves about permutations and combinations, which we met in Chapter 11.

To be able to calculate probability you need to know the number of different ways your desired outcome can happen, and the total number of possible outcomes. Then you can describe it mathematically as

$$probability\,of\,desired\,outcome = \frac{number\,of\,ways\,it\,could\,happen}{number\,of\,total\,possible\,outcomes}$$

Combinations and permutations are related to this: they are about making a selection from a set of possible outcomes and the number of ways this can be done. With combinations, the order doesn't matter, and with permutations, it does. So for the numbers 123, there is one combination and (if there are no repetitions) six permutations ($1 \times 2 \times 3 = 6$). We can then say that if we have n different things, there are $n!$ permutations (the ! sign meaning factorial: $5! = 5 \times 4 \times 3 \times 2 \times 1$). As we saw earlier, Gersonides had worked this out with permutations in the fourteenth century, but we don't know whether Pascal knew of this or discovered it himself.

Let's look at an example of permutations in probability. Say you have five hats of different colours, and you choose three of them. You want to work out all the total number of ways you could do this (getting red then green then blue, or blue then yellow then red, and so on). Once you've chosen a hat you can't choose it again (no repetition is allowed, and the order in which the hats are taken matters). The total number would be $\frac{5!}{(5-3)!} = 60$. There are sixty permutations of picking three things out of five.

Pascal showed that you could do a similar thing with combinations – using them in a completely new way, to solve problems of probability. Let's look at the hat example again. We want to find out the total combinations of picking three hats out of the set of five (this time the order doesn't matter, though there is still no repetition). That would be $\frac{5!}{(5-3)!} \times \frac{1}{31} = 10$. If we adopted some letters for that, we could say that to find the number of combinations of taking k things out of the set of n things would give (after a little tidying up) this formula: $\frac{n!}{k!(n-k)!}$.

Now we come to that genius thing that Pascal did, and why he still probably deserves his name on the triangle. Let's continue with our example of picking three things out of five. Pascal said that to find the number of combinations for this, look at the row beginning with 1 and 5, then go to the third cell in that row. That gives you the number of combinations, which is 10. Easy! He also used this triangle to solve the problem of the unfinished game. The division of the pot between the two players was based on finding the ratio between the sum of all possible (and calculated) combinations, and the total number of all combinations. Pascal's triangle speeded up the process of calculating numbers of combinations and grouping them, rather than listing them all, which would be very tedious indeed.

Probability theory was born, and it flourished amid the 'new science' of the seventeenth-century Scientific Revolution. We're so used to the idea of probability that it's hard to imagine how ground-breaking this concept was to thinkers at the time. As soon as others heard of it, they got very interested. Doubtless the most important of these was Christianus Huygens (1629–95), who was interested also in analytic geometry, but we mention him here for his book *Libellus de Ratiociniis in Ludo Aleae* (*The Value of All Chances in Games of Fortune*). This became a standard text on probability and games of chance. But such games were just an easy, shorthand way of conceptualising what would become a much more important tool in applied mathematics. Through these principles people were now able to calculate probabilities of other events taking place.

We've already met Pascal in an earlier chapter, but we must say a little more about Fermat, who is often credited with being the founder of the modern theory of numbers. His name is most often connected with *Fermat's last theorem*. It's called that because it only became publicly known in 1670, long after his death, when his son noticed some of his father's scribbles in his annotated copy of Diophantus' *Arithmetica*; Fermat had stated the theorem, adding that he had a proof, but it was too long to fit in the margin! Fermat's last theorem is related to Pythagoras' theorem. The formula is $x^n + y^n = z^n$, and it says that you can't find three numbers x, y and z such that the sum of two cubes would equal the cube of the third number; it is valid only for $n>0$, equal to or smaller than 2. Fermat's last theorem ignited the efforts of mathematicians worldwide to prove it, efforts which lasted for centuries; as we shall see in Chapter 36, it was over 300 years later that Andrew Wiles finally came up with a very, very long proof.

Fermat discovered many other interesting mathematical morsels. Fermat's little theorem, for example, says that any integer raised to the power of a prime number will result in a number that the prime number can be evenly divided into without leaving a remainder. So Fermat, like Pascal, wasn't only known for the birth of probability, although that would be an achievement big enough for most people.

Working out probability is always a little tricky as it inevitably involves some element of guesswork, although we base it on mathematical calculations. There are two main ways of looking at probability: subjective probability and frequentist probability. Subjective probability involves making a numerical estimate based on your opinion or knowledge of some event taking place. So you might say your favourite football team is 80% sure to win their next game as their star striker is back from injury. Frequentist probability is where the calculation is based objectively on the long-term frequency of a particular outcome, and we find it as a ratio between the possible desired outcomes and the total number of all possible outcomes. So in that case, you would look back at your team's record and divide the total number of games they have won by the total they have played, which would give you a probability of the actual outcome. Of course, that might still be wrong!

Why is the work on probability important? Practically, we couldn't live without it in our modern world of complex interacting systems. We are talking here about the development of the science of risk, or actuarial science – not something that seemed an obvious pathway when probability was first coming onto the mathematical scene. From the late seventeenth century onwards, actuarial science applied mathematical and statistical methods to assess risk, and today is crucial to the insurance industry, financial markets, investment funds, aviation, medicine, just about anything where predicting whether something you identify can happen, and calculating it with mathematical precision so you could make good decisions.

From a few letters shared between two brilliant mathematicians, an entire branch of mathematics was founded and a whole new perspective on measuring chance and risk became possible. But there was one Englishman who cared not a fig about Pascal and Fermat's card-game solution because he would brook no interruption to his card games. In 1762, it is said, he refused to break off a game when dinner was called, asking instead for his roast meat to be placed between two slices of bread and brought to him so he could eat at the table and carry on playing. Thus, the Earl of Sandwich birthed a new dish and a new way of eating – and avoided having to calculate how to split the pot using the remarkable new mathematics of probability.

CHAPTER 19

Closed Doors and Open Minds

Most mathematical practices, wherever they are invented, are in some way and to some degree made in conjunction with or in opposition to those from other cultures. And we've already seen instances of when this went further, when some mathematicians actually corresponded and even met with those in other countries to discuss and develop their ideas. This sort of cross-border cooperation would only increase as time went on, and characterises a lot of the way mathematics is done today.

But in one country in this period, international contact was strictly forbidden. After a century-long civil war, Japan's borders were firmly closed, and the country entered a period of isolation, later called *Sakoku* (1603–1868), meaning that all foreign influences – including any communications in the sciences, mathematics too – were eliminated. The mathematics, and everything else, being done outside of Japan became unreachable to the majority of the Japanese population. And behind its closed doors, a very distinctive and very beautiful new mathematics was cultivated, with ideas that were different from most of those being pursued by the rest of the world. There is even a distinct name for the Japanese tradition of mathematics from this period: we call it *wasan*, from *wa* (Japanese) and *san* (calculation).

Wasan was not completely devoid of outside influences. The abacus, invented in mainland China, had already been introduced to Japan before the Sakoku period, and mathematics using the Japanese version called the *soroban* was developed in the seventeenth century. A diary from the 1650s written by someone living in the province of Kawachi (present-day Osaka) describes how his generation had already very much developed the practice of using the soroban. Before his time, he complained, people had used rods to compute with, which he found rather inefficient. For more difficult work, such as solving quadratic or cubic equations, some rods were still used, called *tengenjutsu*. Sometimes this word also refers to fortune-telling practices in Japan, but the mathematical method involved positioning rods to represent an equation, drawing from the Chinese technique also imported to Japan. The soroban itself was not that easy to use unless there was someone knowledgeable on hand to show novices how to calculate with it. It spurred into life a whole industry of how-to books. *Jinkoki*, the first such manual explaining how to use soroban, was written by the Japanese mathematician Yoshida Mitsuyoshi (1598–1672) in 1627.

Seki Takakazu (c.1642–1708) was a luminary figure in wasan. His contributions were wide-ranging. He was a contemporary of Isaac Newton – indeed he has been described as 'the Japanese Newton' – and developed mathematical methods for the same purpose for which calculus was created (we'll see more on this in the next chapter). The conceptual foundations for Seki's method, which was called *enri*, were however very different. Seki created algebraic notation for enri and is further credited with a number of discoveries which he developed independently from the European mathematicians of the time. One of his discoveries was related to numbers that we now call Bernoulli numbers.

Unbeknownst to each other, Jacob Bernoulli (1655–1705) and Seki Takakazu were working on the same problem at the same time.

They were trying to understand sequences and series of integer powers – the whole-number powers. Take for example the second power, or square numbers. A sequence of integer squares is as follows: $1^2, 2^2, 3^2, 4^2 \ldots n^2$ (which means we are looking at an infinite sequence of such numbers). There are different formulae you can use to sum such sequences (square numbers, cube numbers, and so on), and the formulae were different for different powers.

Many mathematicians before Seki or Bernoulli had tried to come up with a general formula that would work for any such series in this form – $1^k + 2^k + 3^k + \ldots n^k$, where k is a positive integer – for different powers: cubes, fourth powers and so on. Before Seki and Bernoulli came up with their numbers, though, if people wanted to calculate different sums for different integer powers (square, cube, fourth power and so on), they looked to use different formulae. Bernoulli numbers changed that. They are constants that provide us with uniform ways to find sums of all series of integer powers. In other words, people could now use just one formula and substitute in the appropriate Bernoulli number to find the sum of an integer sequence of a given power.

Seki, in his book *Katsuyō Sanpō* (*Compendium of Mathematics*), posthumously published in 1712, described such numbers using a different method and different notation to that used by Jacob Bernoulli, whose own result on these numbers was published in his book *Ars Conjectandi* (*The Art of Conjecturing*) in 1713. So Seki was actually the first to discover the Bernoulli numbers. Why aren't they called Seki numbers? Well, his results were not known outside of Japan for some time.

Not much is known of Seki's life, but he had been born into the samurai class and came to take up official posts in government, first as examiner of accounts for his local lord, and then as master of ceremonies for the *shōgun* himself, a position of honour. Seki became a samurai, a member of the highest social class in this period, and also a military elite, in the shōgun's clan. He published only one book during his lifetime, *Hatsubi Sampo* (*Mathematical Methods for Finding Details*), in 1674, in which he solved fifteen problems which had been posed four years earlier. He left behind many mathematical manuscripts, and his disciples later published books under his name. This is another important contribution to mathematics with which Seki is now credited: he was particularly notable for inspiring others to do wasan. He formed a mathematical circle where his disciples gathered around him and from which sprung private schools around the country. Some samurai joined Seki's circle, which helped to spread mathematical learning through the country. This kind of learning was not unique to wasan; it was also the way other things were invented, learnt and disseminated. The highly refined arts of *sado* (tea ceremony), *kado* (flower arrangement) and the poetic style of *haiku* were all nurtured in such circles, based on a model of a master and his disciples.

One of Seki's followers was Takebe Katahiro (also known as Takebe Kenkō, 1664–1739). As Seki's student, Takebe came into contact with the ruling families and also served the shōgun's household. Takebe commented on Seki's work, made it his mission to disseminate it as widely as he could, and built on it himself. He developed enri further. This method, parallel to that of calculus 'proper', also led Takebe to discover a number of things independently and earlier than European mathematicians. Takebe's discoveries were linked to the trigonometric functions (those appearing in a right-angled triangle): the sine (*sin*) of an angle (labelled θ) is a ratio of the side opposite the angle and the triangle's hypotenuse; the cosine (*cos*) is the ratio of the adjacent side and the hypotenuse; and the tangent (*tan*) is the ratio between the opposite and adjacent side (or between sine and cosine).

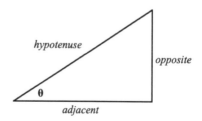

Takebe created the first trigonometric tables in Japan, and published charts of the trigonometric functions. Takebe also did approximations of π – the ratio between the circumference and diameter (or its half, the radius) of a circle, as we know – finding a value correct to forty places. He also played a major part in the publications of the twenty-volume *Taisei sankei* (*Comprehensive Classic of Mathematics*, 1710), a major project that had been initiated by Seki in 1683 as part of his drive to spread the knowledge of mathematics. Takebe wrote down the first twelve volumes following his master's findings and developing them himself; the final eight books were his alone. Even today, Seki and Takebe are celebrated in Japan as the two most distinguished mathematicians from this period.

It may well have been the wasan schools' culture of posing problems to each other that gave us some of the most beautiful artefacts in Japanese mathematics from this period. The *sangaku* (meaning 'calculation tablet') were wooden plaques hung at the entrance to Shinto (and sometimes also Buddhist) temples. Sometimes colourfully painted, these plaques were inscribed with strange, difficult geometrical puzzles – questions, sometimes partial or with possible answers, and diagrams not only of great and intricate visual beauty, but of very interesting mathematics too. The problems were written in Kanbun, a form of Chinese used in Japan until the middle of the twentieth century, usually for official and intellectual works. This was not the writing known to everyone at the time, so this in itself indicates that the diagrams had a limited audience. Around 900 tablets from the period have survived; the total made during the Edo period was probably somewhere between 1,800 to 5,000. Considering the million souls who lived in Edo (today's Tokyo) in the mid-eighteenth century, that's not a huge number. These were tablets made for the educated, by the educated. Were they placed at the temple entrances as offerings to the gods, to organise wasan competitions between mathematical groups, to publicise the cleverness of sangaku masters, to challenge mathematician-wannabes to solve the questions, or to inspire visitors to the holy sites with a quasi-religious contemplation on the order and beauty of geometry? The truth likely involves some or all of these aspects.

In time, the sangaku from across Japan were copied and collected. The prominent mathematician Fujita Kagen (1765–1821) wrote the first book on these in 1790, the *Simpeki Sampo* (*Mathematical Problems Suspended Before the Temple*), where he described the tablets, how he came across them, and explained ways to solve the posed questions. The mathematician Kazan Yamaguchi (dates unknown) owned a copy as a young man, and was later inspired to make his own trip around Japan, visiting every shrine to record the sangaku he found, investigate the problems and offer some solutions. He undertook six pilgrimages between 1817 and 1829, recording eighty-seven sangaku problems in his diary. Only two of these survive. One of them is now called the Japanese theorem. We start with a circle and a cyclic polygon (one whose vertices are placed on the circle's circumference). The Japanese theorem states that no matter how you divide a cyclic polygon into triangles, the sum of the radii of the circles you can inscribe in them will always be constant. It is an amazing and beautiful fact.

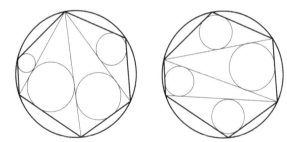

Of all the European nations, the Dutch held a privileged position in Japan during the Edo period (1603–1868). Until 1853, they were the only Europeans allowed in the country, operating from the island of Dejima in Nagasaki Bay, and with the Dutch trading monopoly came cultural exchange. The general ban on foreign books was partially lifted for those in Dutch in 1720, permitting a flow of scholarship and knowledge into Japan which the Japanese called *rangaku* ('Dutch learning'). Through this opening up of communication channels, knowledge of a great many innovations from the West was able to enter Japanese intellectual culture. The 1787 *Kōmō Zatsuwa* (*Sayings of the Dutch*; literally, and rather wonderfully, 'red-hair chitchat'), for example, described many Western inventions, from hot-air balloons to medical developments. In the mathematical sciences, the influence was more of an applied nature. Astronomy and music, as well as mechanics, optics and particularly cartography, as they were then known and developed by the Dutch, became understood in Japan through rangaku, paving the way for the country's rapid modernisation after it was forcibly opened to foreign trade again in the nineteenth century.

Japanese mathematics from this period gave us the fascinating sangaku and a particular way of communicating and meditating upon mathematical knowledge. There is still a steady stream of mathematics enthusiasts around the world who are followers of this practice. Through those exquisite icons of mathematical art, we have a stunning visual understanding of mathematical principles from the period, and a material manifestation of the ways of thinking prevalent in Japan at the time.

CHAPTER 20

The Lion and the Witch

In 1696, Johann Bernoulli (1667–1748) posed a mathematical problem to the readers of a mathematics journal co-founded by the famous German mathematician Gottfried Leibniz (1646–1716). The problem was addressed to 'the most brilliant mathematicians in the world' and challenged them to work out what was the curve of the fastest descent.

What is the fastest path between two points? The *shortest* is the straight line, but as it turns out, this is not the *fastest* way. Though it seems counterintuitive, if you rolled balls down a straight slope and a curved slope, starting and ending at the same points, the quickest descent will be along the curved line, despite it being longer.

Bernoulli had important reasons for posing this problem. He wanted to solve the problem of the path of fastest descent, that is true; he had some ideas about it himself, and wanted to see what could be learnt from the solutions others found. But he had a much bigger plan in mind. Bernoulli and Leibniz were sure that only someone who knew calculus could solve this problem. And in the

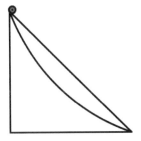

midst of a huge controversy raging at this time, he wanted to show once and for all that it was his mathematical mentor, Leibniz, who was *the* inventor of calculus. Bernoulli was hoping that the other man in this fight, Isaac Newton (1642–1727), then the most eminent scientist and mathematician in England and rival claimant to have invented calculus first, would not be able to solve this problem, granting Leibniz the victory.

Five replies from leading mathematicians duly arrived, offering solutions. One of these was, of course, from Leibniz, and one was, mysteriously, submitted anonymously. Against the author's hopes, Bernoulli recognised the person who had written it, exclaiming in desperation, 'I know the lion by his paws!' Indeed, it was the work of none other than that lionlike mathematical prodigy, Isaac Newton.

The heated controversy over the invention of calculus is a famous episode in the history of mathematics. To see how it all started, we need to go back a bit, to when Newton and Leibniz first started working on it.

When Newton started his university degree at Cambridge, the plague broke. It was 1665 and like many others, the twenty-two-year-old Newton retreated to his home, Woolsthorpe Manor. During the next two years, while virtually a recluse, he dedicated himself to his studies, coming up with the core ideas of much of what he would later become famous for. He taught himself mathematics from the works of the mathematical masters who came before him. Later, in a letter to a friend, he said that 'if I have seen further, it is by standing on the shoulders of Giants'.

And what Newton did see. His accomplishments were staggering. He came up with many crucial, pathbreaking discoveries in mathematics, from intricate problems of areas, volumes, rates and ratios which had plagued mathematics since the ancient Greeks, to solving the age-old problem of planetary motion. Two of his inventions are of particular note – and both have something to do with the problem of the line of fastest descent.

One is, obviously, the apple incident. The story goes that Newton was sitting under the apple tree in his garden at Woolsthorpe when – bonk! – an apple fell on his head. (We now know it didn't fall on him, but there was a falling apple nevertheless.) This prompted him to think about why and how apples fall in the first place, and what is the force by which they do – is it dependent on the size of the apple? From thinking about this force that makes every object fall to the ground, he developed his theory of universal gravitation, which tells us that every particle attracts every other with a force that is proportional to the product of their masses. He explained this in one of the most important books in the history of mathematics as well as science: *Philosophiae Naturalis Principia Mathematica* (*Mathematical Principles of Natural Philosophy*). This was published in 1687, and it contained some mathematics that was not really very easy to understand, but which added fuel to the fire of the controversy later.

The other most important of his inventions was calculus. Newton started working on this at Woolsthorpe in 1666. He began by looking at the coordinate geometry of Descartes (Chapter 17) and studying the curves that could be generated by movement, as explained by van Schooten (Chapter 16). Calculus is a way of understanding and calculating the mathematics of change. To start off with, let's understand that we can describe a dependence between x and y variables by an equation, and draw the line of that equation. Now imagine that line is being generated by the movement of a point that leaves a trace. If you stop at any time, you can draw a tangent to the line, at that point. Newton worked this out through imagining the movement between points that are infinitely close to each other. Every line is composed of a string of infinitely small points, indivisible quantities. They can't be divided any further. Distances between them are infinitely small. But how big is infinitely small? At which stage of dividing the line into the smallest distances can we say that we have got to the smallest, further-indivisible quantity? We could argue about that for a very long time, and people had done so for two millennia.

Instead of getting stuck on such (literally) minor details, Newton forged ahead. He named those smallest quantities *indivisibles*. He called his variables *fluents*, and as they changed, they 'fluxed' through time and moved. The rate of change of a fluent he called a *fluxion*. The fluxion in Newton's language refers to finding the *derivative* of a curve – the derivative is the result of a process of calculus called *differentiation*. Its value will be exactly equal to the value of the gradient of a straight line at the point we are looking at. The gradient is calculated as a ratio of the differences of y and x for any given point. These differences are infinitely small, but we can imagine a much-enlarged example of what is going on here with a fluxion looking like this:

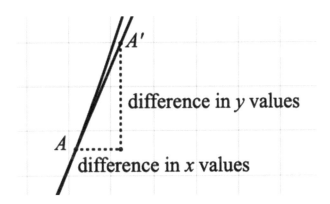

Newton's invention of differentiation set things in motion, so to speak. Using this procedure you could work out the gradient of the line at any point, and from there construct a tangent to the curve you were studying. So now it was possible to calculate instantaneous rates of change between two variables, through this new thing, called calculus.

Over in Germany, Leibniz started working on his ideas of calculus in 1674. He took a different perspective. Our moving point, to which we can draw a tangent from its stopping points, can therefore be imagined to stop at the minimum or maximum values of our curve. These extreme values are seen as the values y can take. Now this may not seem like a big deal, but it was. By being able to find the tangent to a curve at any point, you can work out whether your line will go up or down, and whether at this point your line is at a minimum or maximum value of y.

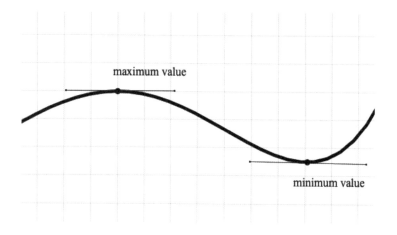

Almost unbelievably, Leibniz and Newton had discovered this revolutionary method completely independently of each other. Leibniz was a giant of philosophy, law and science, a well-connected and fascinating man with many common areas of interest with Newton. Within a decade he was ready to publish the results of his work on calculus. *Nova Methodus pro Maximis et Minimis* (*New Method of Maximum and Minimum*) was published in 1684 in *Acta Eruditorum*, the first journal dedicated to mathematics and science which, handily, Leibniz himself had been involved in establishing a few years earlier. This was also the journal where Bernoulli posed that infamous problem on the line of fastest descent in 1696, from which ensued probably the greatest recorded misunderstanding in the history of mathematics.

A French mathematician, Guillaume de l'Hôpital (1661–1704), now weighed in on the problem. He had read Leibniz's paper from 1684 and he had read Newton's *Principia* from 1687. The same year that Bernoulli's problem appeared, 1696, l'Hôpital had published a piece on Leibniz's invention saying that Newton's *Principia* was 'nearly all about this calculus', just from a different viewpoint. Newton had only published something exclusively devoted to calculus in 1693 (and would again, more fully, in 1704); he was famously reluctant to publish his work until forced to by his friends – or his enemies. So here was the discrepancy. Leibniz had published in 1684, Newton in 1687 and more definitively in 1693. Who was the first to have come up with calculus?

And so commenced a fierce battle of wills. Newton accused Leibniz of plagiarising the unpublished ideas he had been working on since 1666. When Bernoulli's problem appeared, Newton was absolutely outraged. At the time he was Master of the Mint, protecting the nation's coinage from counterfeiters. Right at that moment he was tackling a very successful counterfeiter and had to recall and remake all the existing coins across the English kingdom. To be required, at this time of high stress, to defend his mathematical invention was a step too far. He felt he was being teased by the Continental mathematicians, and this was unacceptable. He brought the fight to the Royal Society, of which he was then president. He found huge support among the mathematician members, and they came out overwhelmingly in his favour. The English-speaking world agreed that Newton was the inventor of calculus. The Germans continued to celebrate Leibniz as the inventor, though he died not long after in 1716.

It is a great shame that two of the greatest mathematicians of the seventeenth century spent their final years fighting over something that they both contributed to. Today we accept that both men invented and described calculus independently, coming at it from their own unique viewpoints. But how was the problem that kicked off this huge row actually solved?

Five mathematicians submitted solutions to Bernoulli's problem: Leibniz and Newton, as we know; l'Hôpital, who was at the middle of the controversy; Johann's brother Jacob Bernoulli; and German mathematician Ehrenfried von Tschirnhaus (1651–1708). They employed different methods, but generally they used the curvature at each point, and by finding the gradient of the tangent to each point were able to prove that the speed of the ball going on this curve is faster than if it went on a straight line. The curve of the fastest descent is called a *brachistochrone* curve, which in ancient Greek means 'shortest time'. It is a cycloid, and can be constructed as follows. Imagine a circle rolling along a straight line. Now imagine that there is a fixed point on that circle's circumference; as the circle moves, that point will move along with it. This point will trace a curve, and that curve is a cycloid. Next time you cycle, your bicycle wheels will be making many imaginary cycloids.

The Leibniz–Newton calculus controversy rumbled on long after the mathematicians' deaths. The prevailing sentiment in the eighteenth and early nineteenth centuries was for Newton in England, whereas mathematicians on the Continent were predominantly on the side of Leibniz.

There were pockets of Newton supporters in Europe, though, especially in mathematical circles in France and Italy. Two followers of Newton's method were women: Gabrielle-Émilie Le Tonnelier de Breteuil, marquise du Châtelet (1706–49), who translated Newton's work into French; and Maria Gaetana Agnesi (1718–99), who translated it into Italian. They were important not only for making Newton's work available in their native languages, though. Breteuil and Agnesi were really the first serious female mathematicians in Europe since Hypatia.

Agnesi was particularly forceful and interesting. She argued for women's right to higher education and was given a chair of mathematics at the University of Bologna, an appointment made by none other than Pope Benedict XIV. She never took it up, but officially she was the first female professor of mathematics at a European university.

She published her own interpretation of Newton's calculus as *Instituzioni analitiche ad uso della giovent Italiana* (*Foundations of Analysis for the Use of Italian Youth*) in 1748. It was an original contribution to calculus in what became known as the Newtonian tradition. Here she gave an example of a curve she called *versiera* (in Italian, 'that which turns'). This curve was already known to some mathematicians: Fermat mentioned it, and another Italian mathematician had constructed it earlier in the century. But Agnesi's book popularised it in ways she couldn't have foreseen.

Recognising that this was a good introduction to calculus and that young mathematicians could learn well from it, Agnesi's book was translated into English in 1801. But its translator, mathematician John Colson (1680–1760), made a mistake. He took 'la versiera' to refer to 'l'avversiera', which in fact means 'the witch'! And so, ever since, we've been calling this curve *the witch of Agnesi*.

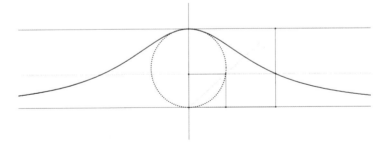

Calculus gave us tools to study and analyse curves better, by finding their extreme values. Now we can use calculus to work out a tangent at any point in a process of differentiation. We can also find the area under a curve by looking at those infinitely small values between points, constructing little shapes between them and the curve, then adding them all up. This is called *integration*. It turns out that differentiation (the part of calculus that deals with finding tangents to the curve at specific points) is the opposite of integration (which is finding the area that a curve makes with coordinate axes), and vice versa. Witchy stuff indeed.

CHAPTER 21

The Travelling Mathematician

Leonhard Euler (1707–83) had been sent to the University of Basel to train to become a Protestant minister, just like his father. But it didn't take him long to realise that mathematics was his true calling. With none other than Johann Bernoulli, his father's old friend, to tutor him, the young man's talent for mathematics was recognised and he switched courses. He became friends with two of Johann's sons, Daniel (1700–82) and Nicolaus II (1695–1726), who were also mathematicians.

About 1,500 miles away from Basel, something exciting was happening. Peter the Great, Russia's first emperor, had grand plans. He established a new city in his name, and there he founded the St Petersburg State University and, with Leibniz's guidance, the St Petersburg Academy of Sciences in 1725. He invited foreign mathematicians and engineers into this new home of science where they were welcomed as professors to share their knowledge. An invitation was extended to Daniel and Nicolaus II and they entered the service of St Petersburg Academy that year. When Nicolaus II died a year later, Euler replaced him.

This was Euler's first major journey, and he was to stay in St Petersburg for the next thirteen years. When he arrived there, he met a Prussian mathematician who had come to the academy upon its founding. Christian Goldbach (1690–1764) and Euler remained friends and correspondents to the ends of their lives.

The two men discussed a curious problem centred on Goldbach's hometown of Königsberg, a city that occupied both sides of the River Pregel and two large islands, which were all connected together by seven bridges. Was it possible in a single trip to visit all four land masses, crossing each bridge only once? Euler thought this was really a problem of logic, but nonetheless sat down to solve it mathematically. He even corresponded with the mayor of Königsberg about it. Though he didn't leave St Petersburg, and travelled there only in his imagination, this was to be the second major journey of Euler's life.

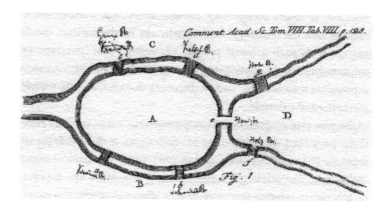

Very quickly, Euler realised that concentrating on the route within the land masses was unimportant. The problem was related to the sequence of bridges crossed. He explained that for a journey to begin at one land mass and end at another, then there would need to be an *odd* number of connecting bridges. But each land mass, except the two where the journey is started and ended, needed to have an *even* number of bridges connected to it. In the Königsberg problem, each of the four land masses is touched by an odd number of bridges, so there was a contradiction – the proposed path was impossible.

Euler's radical reformulation of the problem in such abstract terms laid the foundations of *graph theory*. We can draw a diagram of the Königsberg bridges problem, where the land masses are replaced by points (vertices, or nodes) and the connecting bridges by lines (arcs, or edges), in what is now known as a *planar graph*. Here we can see clearly that every vertex had an odd number of arcs attached to it. And because the journey over the bridges of Königsberg has to cover all the edges, the vertices with an odd number of arcs would have to be at the beginning or end of the journey to make the route possible.

Euler generalised this for all such journeys. As a consequence, we have something we call an *Eulerian path*, which is a continuous path that passes through every arc once and once only. A related theorem, also by Euler, states that if a network has two (or fewer) odd vertices, it has at least one Eulerian path.

Interestingly, were you to visit Königsberg (or rather, Kaliningrad) now, you *could* cross all the bridges by following an Eulerian path. Only five bridges remain following the Second World War; two land masses now have two bridges, and the other two have three bridges, so it is possible to make the journey, but only if you start and end on different land masses. In other words, it is not a *Eulerian cycle* (a Eulerian path that starts and ends at the same point). We can make the journey now because the *topology* of the city has changed. And topology was indeed a new branch of mathematics created in Euler's day. It deals with the way geometrical objects are made – how their edges are connected, and other such characteristics – but doesn't deal with the measurements of things. A topologist would ask questions such as 'How do the edges connect?' or 'Does this thing have holes in it and how many of them are there?' rather than 'How long is their connection?' or 'How big are the holes?' This would become an increasingly important branch of mathematics going into the twentieth century.

What Euler discovered at the end of his imaginary journey has become his most famous contribution to mathematics, but it was one of his many other inventions. He was astonishingly productive; by some accounts he must have written a paper a day for the majority of his adult life. He also came up with so much new mathematical notation that it is difficult to choose which things to mention.

Perhaps the most important is the number e, or Euler's number, a numerical constant used in mathematical calculations, a bit like π. Such numbers describe some important relationships, but you can't find their exact value, only an approximation. The number e starts like this: 2.7182181828459045235 . . ., but since it is also an irrational number it doesn't end, and so it can't be written as a finite fraction. That is why it is useful to have an e instead of writing all these digits. Some say Euler named the number e after himself! But he never mentioned anything of the sort.

The number e was known but had no name until Euler came along. It is the base of natural logarithms. Remember that we saw logarithms with base 10 in Chapter 16. That logarithm could be written as $log_{10} a = b$, and would mean logarithm, base 10, of a is equal to some b. We also know that this means the same as saying that 10 to some power b is equal to $a - 10^b = a$ – it's just written in a different way. Now Euler said you can have logarithms with the base e. That would be written like log_e, but instead of having to write e all the time, you could also just write log slightly differently, which is what we do now; we write logarithm of base e as ln.

Until Euler started looking at the number *e*, people didn't really know how to calculate it entirely. The value itself came out of trying to calculate compound interest, which is an important concept in lending, borrowing or saving money. It means that you incur interest on the original loan and also on the previous interest which is added to the loan.

Jacob Bernoulli, whom we met in Chapter 19 and who was the brother of Euler's teacher Johann, was working on this. He came across a formula which worked for different periods of compounding. For example, say you have £1, and your interest is 100% calculated once a year. This would yield £2 at the end of that year. If the interest is compounded twice a year (which means that 50% interest is applied twice a year rather than just 100% applied at the end), it will give you $\left(1+\frac{1}{2}\right)^2 = £2.25$. In general, you could calculate it as $\left(1+\frac{1}{n}\right)^n$, where *n* is the number of times the interest is compounded. As you increase the number of periods (as *n* becomes larger), the result is always bigger than the previous value; it is more than £2, but also less than £3. And this is as far as Bernoulli got. He realised that this number, later to be identified as *e*, is some kind of constant, the value of $\left(1+\frac{1}{n}\right)^n$ as *n* goes to infinitely large values.

This brings us to the new concept, that of a *limit*. A limit is a certain value that a number sequence or series *tends towards*, but never quite reaches. A famous example of this is called the *Basel problem*, so named by Euler, who solved it after a host of top mathematicians had been defeated by it. This concerns the limit of the infinite sum of reciprocal squares. The reciprocal of 2 is $\frac{1}{2}$, so it was the sum of infinitely many fractions like this: $\frac{1}{1^2}+\frac{1}{2^2}+\frac{1}{3^2}+\frac{1}{4^2}\ldots$

Euler calculated that this is approximately equal to $\frac{\pi^2}{6}$. How do you get to that value? After many long hours of calculations and seeing what results you get as you increase your sum. Euler used the same trick for finding a value for *e*. If he could find a sum of a number series that tends towards the value of *e*, all would be well and good. He succeeded and found that *e* is exactly equal to a sum of reciprocals of factorials of natural numbers: $e=1+\frac{1}{1!}+\frac{1}{2!}+\frac{1}{3!}+\frac{1}{4!}\ldots$

There were other things Euler found and invented. Because he did so very many sums like this he came up with a sign for summation, using the Greek letter sigma, so that the sum of some numbers n can be written as $\sum_{n=1}^{N} n$, which would mean $1 + 2 + 3 + 4 + \ldots + n$. You read this as 'the sum of n numbers going from 1 to n'. This must have been useful as it sped up the writing down of all the sums Euler was working on. With Euler, mathematics gained many shortcuts.

In 1740, Euler left St Petersburg for Berlin, then the capital of Prussia – his third major journey. By this time he was having real problems with his eyesight, and he hoped to find a kinder climate than that in Russia. He became tutor to the niece of Frederick the Great, King of Prussia. Working for a king meant that you were really regarded as a great mathematician, and Euler had found himself twice in that position. All the while, he remained in touch with his good friends and students in Russia.

But even Euler couldn't solve every problem. His best friend Goldbach wrote to him regularly, and in one letter posed a very tricky problem, now called Goldbach's conjecture. He suggested that every even integer greater than 2 can be represented as a sum of two primes: $n = p_1 + p_2$. Now, it's important here to remember that at that time, by convention, 1 was considered to be a prime number (it is not now). Euler said that he was certain that what Goldbach had written was true, but he couldn't prove it. And no one else since has been able to prove or disprove it. This is one of the famous unsolved problems in mathematics.

In 1760, during the Seven Years' War, Euler's grand house near Berlin was attacked and ransacked, and he returned to Russia in 1766. This would be his final journey. His eyesight deteriorated even further and he became almost entirely blind. But it didn't diminish his energy or his spirit; he is said to have exclaimed, 'Now I will have fewer distractions!' The impetus to do mathematics all day, every day, remained – he simply found scribes to write the new papers he dictated.

The unsurpassable, irrepressible Euler was one of the most productive mathematicians of all time. He is universally known.

Mathematicians celebrate Euler for his amazing output and for contributing so much to mathematics. One of his successors called him the master of all mathematicians. Not being able to solve some things, like Goldbach's conjecture, didn't make him any less important. Sometimes the problems are just not ready for solutions.

CHAPTER 22

Modern Geometry and Its Traces

Imagine a point in space. Space is homogenous; it doesn't have any little quirks here or there, it is just the same throughout. And the point has no length, breadth or width, it is just a little dot. Now imagine this point starting to move slowly, and as it does, it leaves a trace behind. It travels straight in one direction, and as you see it move, its trace will be a straight line. Now imagine this straight line, the trace, is stationary itself. Imagine another straight line crossing over this line in a perpendicular direction, and then that new line starting to move across our first line, it too leaving a trace. This time, the trace will become a flat plane.

This imagining of everything in space being described by a movement of elements and the traces they leave behind was the basis of technique invented by the French mathematician Gaspard Monge (1746–1818). He called it *descriptive geometry*. He perfected this type of description to imagine all space to be generated by such movements and it became an incredibly popular way of teaching geometry in countries around the globe – France, and all the territories where France had influence in the nineteenth century.

Descriptive geometry, you could say, revolutionised both the concept of geometry and the way it was taught. Instead of having a cube as an object, for example, you could imagine it being generated by perpendicular movements of its edges to create a three-dimensional figure consisting of six faces.

The technique required a way of communicating what was going on. You could do that in words, but we have seen that mathematicians always look for shortcuts, symbols or graphs to make things quicker, more precise and universally recognised. In this way, language would not be a barrier to understanding across different cultures.

Because Monge was dealing with space, he used something similar to coordinate geometry. Let's say we have two planes which we will call projection planes. To draw another plane, Monge used a little trick of imagining how that plane would go through these two projection planes: it will *cut* them, leaving a kind of trace on them. We can also draw geometrical objects on these projection/drawing planes as they would be seen from different angles (from the front, and from above). The projection planes could be rotated so one falls into another, and there will be a line to distinguish between them, which is the line where they intersect. This originally highly abstract technique of imagining and drawing objects has now become a standard for technical drawings, but Monge's invention was more than that.

If you say the two planes on which we draw our objects are our coordinate system, we could further connect the image of such an object with its algebraic description. By inventing this system, which we will describe now in more detail, and which borrowed from many things that were already known, Monge is also sometimes known as the father of *differential geometry*. This was a new branch of mathematics using calculus as well as algebra, and usually studies smooth surfaces. Other mathematicians from this period started using similar things in combination, but only Monge used differential geometry with descriptive geometry.

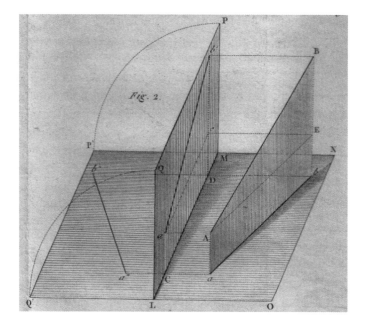

How these are connected seems to have been clear to Monge when he was only about eighteen years of age. He became a draftsman at the École Royale du Génie in northern France, a military college which had only recently been established. This school was founded to train military engineers, and mostly, to start with, in the design of fortifications. These were a crucial element of military strategy at the time. They had to be designed with walls high enough to withstand attacks by cannon and firearms. At that point, the necessary height was calculated by taking measurements of the surrounding terrain, involving heavy work in finding the position of numerous viewpoints. What Monge came up with was to find the highest points, and to then imagine putting tangential planes through such points. This allowed him to work out the required height of the walls: just enough that all the tangential planes to the fortification were met with walls that were a bit higher. As we know, a tangent line is a line that touches an object through a point, and a tangent plane does pretty much the same thing in three dimensions. And if you remember from when we talked about differentiation, we can find a tangent at any point of a curve by finding its differential. Similarly, we can do that to find tangential planes to surfaces.

However simple it seems to us now, no one had yet thought of using this knowledge in the way Monge did. What he called his 'descriptive geometry' method was examined by the highly critical officers of his college, and once they realised this would save them a huge amount of time and be much more precise than their current method, Monge's technique was proclaimed a military secret. It remained so for many years, and was only taught publicly after the French Revolution in 1789. Monge's simply titled *Géométrie descriptive* (*Descriptive Geometry*), based on his lecture notes, was published in 1798. The revolutionary regime founded new schools in Paris, and Monge played a significant role in establishing the École Polytechnique, sometimes also being referred to as its father.

Following this invention, Monge used differential geometry to study smooth curves and surfaces. Let's say you want to find out what the differences are between the surfaces of cylinders and spheres, and measure them mathematically. The curvatures of these two surfaces are different, as the cylinder can be flattened but the sphere can't. If you cut a paper cylinder along one of its straight lines, it will just fold out into a flat plane. You can do the same thing with a cone. But you can't do that with a sphere with one cut only.

Through using descriptive geometry, Monge was able to start studying surfaces in a new way too. This is because descriptive geometry relies on those traces that points and lines 'leave behind' as they move under certain conditions to create curves and surfaces. When we think of mathematical surfaces, we consider their mathematical properties. There is one particular family of such surfaces that is very useful in engineering applications. They are like that cylinder and cone; you can cut them along one line and they will fall flat into a plane. These are called *developable*, and are a small subset of all possible surfaces. Because developable surfaces can be flattened without changes (you don't need to do any creasing, tearing or stretching), any shape on their surface will remain the same once it is flattened – that is, developed into a flat plane. A circle that you draw on a cylinder will be exactly the same when that cylinder is flattened.

Monge was also instrumental in coming up with a problem that has been investigated more recently in the twentieth century. This problem has to be pretty important then, right? It is, but it starts quite modestly. Let's say you have some piles of sand and some holes in the ground somewhere. How do you best move the piles of sand to fill these holes? Once you start working through this problem, you would get many possibilities, or many permutations of possible solutions, that could allow you to achieve the task in different ways. But what is the best way? That would be the one where your route would allow you to fill all the holes from the piles of sand in the shortest possible time and therefore also with the least amount of energy. The problem, in other words, is that of optimisation. Here Monge used the tools of calculus to find minimum values. He discovered the way to find values for the minimal work or minimal transport cost to decide the best way to complete a task. This inspired the development of optimisation theory, which is an incredibly useful tool in transport networks.

Monge was well known in the parts of the world where his technique of descriptive geometry was taught. This did not include many of the English-speaking countries, particularly England itself. This was less about mathematics than it was about politics. Monge had endured a turbulent adult life, having been one of the leaders of the French Revolution, then escaping a death sentence during the Reign of Terror, and eventually becoming a close friend and supporter of Napoleon Bonaparte. France and England were at almost constant war during Napoleon's time, and it was the war that eventually the English won.

Napoleon Bonaparte (1769–1821) is a divisive figure in history, but very few people know that he was a mathematician too. He studied at the prestigious École Militaire in Paris, excelling in mathematics as a student. In 1798 he invited Monge and other mathematicians on his scientific expedition trip to Egypt, where

Monge founded the Institut d'Egypte and wrote an interesting paper on the physics of mirages. One could imagine the two men wiling away the evening hours in the desert discussing the finer points of geometry. In fact they came up with somewhat similar theorems involving three of the same geometrical elements.

Monge's theorem states that, if you have three circles, and none of them are inside the other, then the following always happens. If you draw tangents to pairs of your three circles – and you will have three such pairs – these tangents will all intersect on one line.

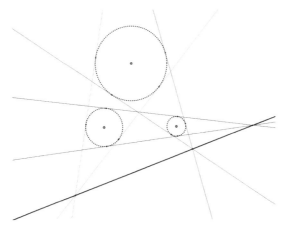

Napoleon's theorem deals with three triangles. It states that if you construct equilateral triangles on the sides of a triangle (any triangle), then their centres will form another equilateral triangle (XYZ). And, if you then construct circumcircles (a circle that encompasses a triangle) of these three triangles, they will intersect at a single point called the Fermat point (F), named after Fermat, who discovered it. The centre of a circumcircle is at the intersection of the bisectors (bisects = cuts equally) of each of the triangle's angles. The Fermat point minimises the sum of the distances from the vertices of our original triangle (A, B and C).

Monge established two grand schools in France, an institute in Egypt, two new branches of geometry, and inspired the creation of optimisation theory. He had many followers of his mathematical work throughout the world, but because of his friendship with

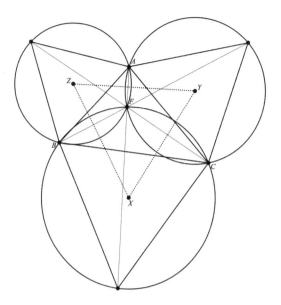

Napoleon, has always been shunned in certain places. He was a loyal friend by all accounts, and was revered for many years after his death by his students from the École Polytechnique where he mainly taught. The area in Paris where this school was first founded has many things named after him even now – a street, a square, a metro station, even shops, pharmacies and restaurants. As one of the 'fathers' of the French Republic (on top of all his other fatherly roles), he is now buried in the crypt of the Panthéon not too far from Rue Monge. It's a moot point between mathematicians whether he was more important as a mathematician or a politician. Perhaps that which remains timeless is more important – let's vote for mathematics!

CHAPTER 23

A New World Out of Nothing

Can a mathematician create new worlds by creating new mathematics? This is what János Bolyai (1802–60) claimed. He wrote a letter to his father in 1823 in which he said, 'I have discovered such wonderful things that I was amazed ... out of nothing I have created a strange new universe.' Whatever could he have meant? To create a new universe, one would have to be a divine being of some kind, or have some never-before-seen power. Or else, an artist who creates an imaginary universe that is nonetheless confined to their works of art. But János was neither god, super-being nor artist. He was a mathematician.

János was born in Cluj (Koloszar), the capital of the Habsburg principality of Transylvania (now in Romania). His father, Farkas (1775–1856), had been born in the nearby village of Bolya (hence their surname), and was also a mathematician. He had studied at the famous German universities of Jena and Göttingen, and at the latter had become a close friend of Carl Friedrich Gauss (1777–1855). After university, Farkas went back to Transylvania where he became a mathematics teacher in a college. Gauss, a mathematician of great stature, became a professor at Göttingen.

János studied there too. He was a child prodigy by all accounts. He taught himself to read at five years old, was first violinist in a local string quartet, and had mastered calculus by the age of thirteen. Aside from his mother tongue of Hungarian, and German and Romanian (languages spoken where he lived), Bolyai taught himself Italian, French, Latin, Chinese and Tibetan. This was a young man of huge promise. Farkas recognised his great interest and talent in mathematics, and János was sent to Vienna to study aged fifteen. His father then asked his friend Gauss, who by this time had become director of the Göttingen Observatory and professor of astronomy there, to help young János by taking him on as a student. But Gauss wasn't helpful. Consequently, János entered the engineering corps of the Austrian army, where he served for ten years.

While he was the army engineer, Bolyai continued working on mathematics. At some point he became interested in the theory of parallels. This was Euclid's fifth postulate, the so-called parallel postulate, which had vexed mathematicians really since the Renaissance when there had been a surge in translations of the *Elements*. The reason mathematicians were so bothered by this postulate was that it looked like something more than a postulate. As we saw in Chapter 5, Euclid's five postulates are the most basic of all statements, meant to be simple, self-evident and not divisible into simpler statements. And here lay the problem that many mathematicians were trying to grapple with over the centuries.

Let's have a closer look at those postulates. The first three deal with what you can do with the Euclidean tools, straight edge (without marking) and compasses.

1. A straight line can be drawn from any point to any other point. Check!
2. A finite straight line can be extended indefinitely in a straight line. Check!
3. A circle can be drawn with a centre and radius. Check! That's pretty much what compasses do.

Now the fourth postulate is a little more suspect, as it states:

4. All right angles are equal to one another.

That seems straightforward, but it assumes quite a few things – that space is homogenous, for example, which has not yet been stated. And this leads onto the final, fifth postulate, so much longer than the others, and even more suspect:

5. If a straight line falling on two straight lines makes the interior angles on the same side less than two right angles, the two straight lines, if produced indefinitely, meet on that side on which are the angles less than the two right angles.

We can sketch that out, exaggerating things a bit. Euclid is imagining something like this: that lines which may even appear parallel, when cut with a perpendicular line, have interior angles just a little less than 90°, and so would actually meet *at infinity*, and would hence not be parallel after all.

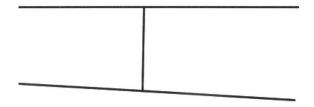

That troubled mathematicians. What happens at infinity? How can we know that? Neither was this postulate very basic – many thought there was too much complexity here for an axiom. So what was left to do? Mathematicians across the centuries and the world tried to simplify it further, to prove that it could be derived from the other four postulates. And some other mathematicians took a different tack. One of the most interesting of those was Gerolamo Saccheri (1667–1733), an Italian Jesuit mathematician. He tried to establish the validity of Euclid by assuming the fifth postulate to be false and seeing where that took him; if Euclid was wrong to formulate this postulate, he said, then there would be a contradiction in the whole system of Euclidean geometry. So, he concluded, Euclid should be 'vindicated'! In this thought experiment, Saccheri actually came tantalisingly close to realising he had discovered some non-Euclidean geometry (which we'll come back to in a bit), but his intentions were firmly to shore up the ancient Greek view.

Try as they might, mathematicians couldn't make any real headway with this. The plucky, brilliant János Bolyai thought he would give it a try. But when Farkas first learnt of his son's intentions to start working on the parallel postulate in 1820, he wrote to him in alarm. He himself had spent many years working on the problem and knew its dangers; he had 'lived through that endless night', he said, and he warned his son, 'It will consume all your leisure, your health, your peace, and all your joy in life.' János was undeterred. By around 1824 he had formulated his theory of parallels, with a proof to show that this postulate was indeed a weak link in Euclidean geometry.

János's approach was ingenious. He went about the problem in a way that was different to anyone else who had attempted it up until then. What he proposed was that we look at the system of geometry for which indeed the fifth postulate is true: he called this geometry the Σ geometry. And then, he proposed, we can say that there is *another* type of geometry, which we can denote as *S* geometry, in which the parallel postulate is *not* true: the parallel postulate doesn't hold. So what this meant was that there was the *old*, Euclidean geometry Σ, but there was now also another, *non-Euclidean* geometry. And János hadn't finished yet. There was an *additional* geometry he called *Absolute*, where that troublesome postulate and everything that stemmed from it doesn't even exist.

In Euclidean space – which is what we're used to – the laws of Euclidean geometry rule. Now let's see what that non-Euclidean space would look like. Imagine a world where parallel lines, when they near or reach infinity, do eventually meet. That would mean that when they are crossed by a line perpendicular to them, the sum of the angles on that line is *less than* 180° on one side. (In such a space, even the term 'perpendicular' wouldn't mean what it does in

Euclidean geometry ... but let's go on.) In another type of non-Euclidean space – which studies curved surfaces, not Euclidean flat planes – a triangle's sides would consequently not be entirely straight, but curved, so the sum of its angles would also be *less than* 180°. Strange things happen in non-Euclidean space!

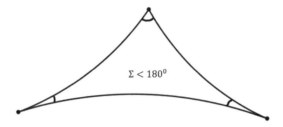

$$\Sigma < 180^0$$

Now, this isn't an easy thing to get your head around – indeed it has bothered mathematicians for centuries. And that is what János meant when he declared he had created a 'strange new world': a world where it is possible to have not only a single mathematical explanation of the parallel postulate, but a few others, and another world where it is possible to have all kinds of other things where parallels don't even exist.

The type of geometry where each line has at least two parallels became known as *hyperbolic geometry*. We have already met with the hyperbola as one of the conic sections, but here we're talking not about a two-dimensional curve, but the surface you get when you rotate a hyperbola around an axis of reflection. We call this surface a *hyperboloid*, and it looks a bit like a saddle. If, then, we take this to be a model of non-Euclidean space, we could imagine everything in it to be slightly curved, like hyperboloid itself.

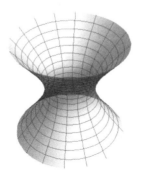

Having worked all this out, János visited his father early in 1825, wanting to convince him of his new theory of geometrical space before he showed it to others. After all, if he was right, all geometrical knowledge up to that point would be put to the test, and a different version of space could be imagined that would still be mathematically sound. Euclidean geometry would become just one of other possible geometries, and a new world, the one János imagined, would be granted the light of day, just as Euclidean geometry had defined the world of geometry for more than two millennia. After visiting his father, though, János was to be disappointed: Farkas was not enthusiastic. But the son was not going to give up that easily. He had a deep belief that this outrageously novel idea of non-Euclidean space, which he had shown was mathematically completely sound, would be recognised by the whole world sooner or later. The question was how to go about getting it out there.

János Bolyai was not, after all, a mathematician – officially at least. He was a military engineer, having entered the army engineering corps in 1823, and was posted around the Austro-Hungarian Empire. On one occasion he discovered the captain was his old mathematics teacher from Vienna, and shared a draft of his new ideas with him, hoping for feedback and, more importantly support. He never got the manuscript back. By chance, on his way to another posting in 1831 he dropped in on his father again. János had ironed out some inconsistencies, and Farkas had had time to properly digest what his son had discovered. He now urged him to publish his work. This appeared in 1832 as an appendix to Farkas's own book on mathematics, *Tentamen juventutem studiosam in elementa matheseos purae* (*An Attempt to Introduce Studious Youths to the Elements of Pure Mathematics*), titled *Appendix scientiam spatii absolute veram exhibens* (*Appendix Demonstrating the Absolute Geometry of Space*).

Farkas sent a copy to his friend Gauss, hoping to get that great mathematician's approval that János had sought for so long. He didn't get it. Though Gauss referred to János in another letter as a 'genius of the first order', his reply to Farkas was frank and galling. 'To praise it would be to praise myself', he said. He had already thought of all this three decades ago, though he hadn't published on it.

Gauss was an extraordinarily productive and important mathematician. It is perhaps unfair to mention him only briefly as he had a huge influence on mathematics in many areas, from number theory – for which he was sometimes called the *princeps mathematicorum*, or the 'prince of mathematics' – to analysis, geometry, astronomy and optics. Perhaps the only blot on his copybook was not supporting János in his hour of discovery and need. But there was truth to what Gauss said. He had certainly been working on non-Euclidean geometry, and as early as 1816 had realised that the fifth postulate couldn't be proved. But the comments stung János, and a niggling sense of injustice gnawed away at him. Had his father mentioned his ideas to Gauss all those years earlier, after János had first come to see him, thus planting the seed of his idea? But János couldn't prove anything. After the greatest of all mathematical appendices was published, there were no offers of employment, no change of fortune, no acknowledgement of his new discovery or the importance of the contribution that János Bolyai made to mathematics. He was a broken man, physically and mentally unwell, and retired to his family estates.

Then, to his great consternation, in 1846 he discovered that yet another mathematician had published his own work focused on parallel lines. His name was Nikolai Ivanovich Lobachevsky (1792–1856), a Russian professor at Kazan University, whose theory on non-Euclidean geometry had appeared in a series of five papers published in the university's journal in 1829. Like that of Bolyai, his geometry was hyperbolic; and also like Bolyai, it had little circulation and attracted very little attention at the time it was written. Gauss only learnt of Lobachevsky from a paper he published in 1837, *Imaginary Geometry*, and corresponded with him for a while. Bolyai, meanwhile, was a rage of emotions. On the one hand he could appreciate the masterful mathematical proofs that Lobachevsky had come up with. But in the next breath he was raving that Lobachevsky didn't actually exist, and that it was all a trick being played on him by Gauss. He was so deeply wounded by not having found support when he most needed it.

It took a good thirty to forty years for all of this to make real impression on other mathematicians. Only in the late 1850s and 1860s did they start to take real note of what both Lobachevsky and Bolayi had done with their work on non-Euclidean geometries. And an English mathematician, Arthur Cayley (1821–95), came up with yet another possibility: elliptic geometry. What's this? Well, let's look again at the parallel postulate and how the different geometries deal with it. In Euclidean geometry, the parallel lines never meet. In an infinitely large space, with infinitely long lines, we can imagine there is more than one parallel line to a given line: that is hyperbolic geometry. In elliptic geometry, there are no parallel lines at all: the parallel postulate is not consistent with its other laws, so it requires a different set of axioms in order to have a consistent system. One of the best ways to imagine parallel lines in elliptic space is to consider the surface of a very large sphere with two lines encircling it drawn on its surface. Looking at the sphere where the circles are far apart, the lines would appear parallel, but taking a different viewpoint, you would be able to see where the lines intersect. Here, then, lines that appear parallel meet twice.

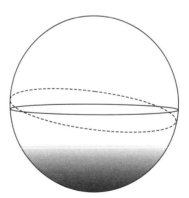

This was indeed a new dawn of revolutionary ideas about space. And these ideas only got more interesting as time went on. Like two parallel lines, the first two inventors of non-Euclidean geometry, Bolyai and Lobachevsky, weren't recognised for their work during their lifetimes, but now every mathematics student knows their names.

CHAPTER 24

Romantic Mathematicians

Ever since Cardano and Tartaglia had solved cubic equations (Chapter 13), mathematicians had been mulling over the problem of solving equations of higher powers (larger than four). No one had been able to come up with a formula to solve equations of degree five and higher in a similar way. Given the importance of this work, you might've thought some highly experienced sage would have worked out the trick eventually, but in fact it was solved by two very young and inexperienced mathematicians.

They lived in the age of Romanticism, when some sweeping changes were taking place across the Western world. People rebelled against the old privileged regimes (as in the French Revolution); colonies rose against those who had established them (as in the American War of Independence). This was also the great age of stories featuring the Romantic hero, a young and lonely figure who rejects established norms and is in turn rejected by society at large. And strangely enough, the story of these two young mathematicians was just such a tale.

The first player on our stage is Niels Henrik Abel (1802–29), a Norwegian mathematician who was very poor, and died aged just twenty-seven. Abel was interested in trying to find some kind of rule for whether equations of higher degrees can be solved. Just a summer before he died, he wrote a short manuscript while on vacation with his fiancée. One can think of more romantic things he could have been doing on holiday! But not Abel. He entitled his work simply *A Theorem.*

That October, Abel submitted his work to the Paris Academy for publication, but alas the manuscript got lost. Rather, it seems that Academy member Augustin-Louis Cauchy (1789–1857), an established and well-regarded French mathematician, put it to one side without properly looking at it. Poor Abel contracted tuberculosis and was soon gravely ill. Without money or employment he couldn't afford a doctor, so his condition rapidly deteriorated. His friend, the German mathematician August Crelle (1780–1855), from a similarly poor background, had established an important mathematical journal. He liked Abel, published some of his work in the journal, and, as an influential mathematician, tried to find him some work.

Abel wrote to Crelle about the *Theorem* he had submitted, and gave a summary of what it was all about. What Abel had found was that there is no solution in radicals to equations of degree five or higher with arbitrary (randomly picked) coefficients. *Radicals* refers to algebraic expressions that involve square roots, like for example $\sqrt{2}$ or $3 + \sqrt{2}$ or similar. But if these higher-degree equations are in some way related to each other, such that one of them may be expressed in terms of the other, then they are soluble in radicals. So, if we pick our coefficients well, we can do it; otherwise we can't.

Crelle did find Abel a job, and Abel's lost *Theorem* was found again, but neither happened before his death. At least some of his thinking had appeared in Crelle's journal; the *Theorem* was finally published in 1841. The discovery is now known as the Abel–Ruffini theorem, to include the Italian mathematician Paolo Ruffini (1765–1822), who started working on it some years before Abel.

Around the same time that Abel was struggling with his health, wealth and work, another young mathematician was contending with his own issues. This was Évariste Galois (1811–32), born into much more comfortable circumstances, who attended the famous Lycée Louis-le-Grand in Paris. Aged fifteen, he enrolled in his first mathematics class and quickly excelled, but other subjects didn't come so naturally, and his school reports are full of comments about how bizarre and troublesome he was. He evidently had great difficulty expressing himself. At school he failed his rhetoric exam and had to repeat a year. His dream was to study at the most prestigious French college, the famous École Polytechnique founded by Monge, but he failed the entry exam twice. Eventually he was accepted into the École Normale, perhaps by a whisker; his literature examiner felt he had answered the questions poorly and seemed to know absolutely nothing.

But while still only seventeen or eighteen, Galois was already writing and publishing mathematics papers of startling originality. In 1829, the year of Abel's death, Galois wrote a paper on exactly the same topic as the Norwegian had written on the previous year, and also sent it to Augustin-Louis Cauchy. Cauchy now told Galois about Abel's work, and Galois duly updated his article, *Mémoire sur les conditions de résolubilité des équations par radicaux* (*On the Condition that an Equation is Soluble by Radicals*), and sent it to the Paris Academy to be entered for the Grand Prize. Unbelievably, after the man he sent it to died, it too was lost! It was never considered for the prize. But here, he had started work on what would become a new theory in mathematics.

All this mathematics was going on against a backdrop of disorder and violence in the French capital. In 1830 France experienced another revolution, overthrowing the monarch Charles X. Rioting in Paris was widespread. Galois was sympathetic to the rioters, and tried (and failed) to scale the walls to join them after the director of the École Normale locked the students in. He was expelled and joined the militia in the Artillery of the National Guard. Galois' political opinions and republican feelings ran high. As such, and being against the establishment, he wasn't able to find permanent work. He was still writing, working on his mathematics, and tutoring, but had no formal position. He was encouraged to send another version of his paper on equations of higher powers to the Academy in 1831. But events overtook him. He made the mistake of publicly threatening the life of the new king at a party of republicans, and was subsequently arrested. He was acquitted but was then again arrested and was in prison when he got the letter of rejection of his paper from the Paris Academy. His ideas weren't clear enough and needed more development. By April 1832 he was out of prison, but just a month later got into trouble again. He challenged someone to a duel, probably related to a woman he was in love with. The night before, he sat down to write his final mathematical words. On 30 May he was wounded in the fight, and died a day later.

In his last manuscript Galois built on the work of Abel. This was the beginning of what we now call *Galois theory* and, although he didn't work all of it out, it gave birth to a whole new area of mathematics. Galois theory is the study of certain groups that are related to those polynomial equations of higher degrees. In short, we know whether or not solutions to equations can be done using radicals, based on the properties of a Galois group.

A *group* is an algebraic structure. It is based on a set of elements over which some *binary* operation (dealing with two objects, like two numbers) is executed. There are groups which use finite or infinite sets. A set of the letters making the title of this book would be {a, l, i, t, e, h, s, o, r, y, f, m, c}, and this is a finite set. There is no need to repeat elements that appear more than once. Or take another, more mathematical example. Let's look at integers. We have an infinite set of those, and for the binary operation we will use addition. We have just created a *group of integers under addition*. This new object – the group – has some really important characteristics.

The first necessary property of a group is that, whatever operation is done to the elements of your set, the results have to also be found within that set of elements. So for our set of integers, once you apply addition, the result will also be an integer: $3 + 4 = 7$. This makes our set *closed* under addition.

The second necessary property is an *identity element*. This element is special. If you apply the group operation on the identity and another element, the result is the original (other) element. In our example, our identity element is 0. When we add it to a number, it doesn't change it: $3 + 0 = 3$.

The third necessary property is that there has to be an *inverse* element for each of the elements in the set. This means that, when the binary operation is done on any element and its inverse, you obtain the identity element: $3 + (-3) = 0$.

Finally, the composition of the operation has to be *associative* – the result you get when applying the operation to three or more numbers doesn't make a difference to the result when they are grouped in different ways. For our example, $3 + (2 + 4) = (3 + 2) + 4$.

There is another property that a group *can* have, which is that the operation is *commutative* – that is, it doesn't matter what order the elements are written in when applying the group operation, you still get the same result. In our example, $3 + 4 = 4 + 3$. If a group has that characteristic too, it is Abelian, named after Abel and in his honour. Our group of integers under addition is therefore an Abelian group.

What was revolutionary about what Galois did was to use such structures to work on equations of higher degrees. Until then, people were just trying to find formulae for solving them. Abel discovered that there is no such formula for equations higher than the fourth degree. But Galois now invented a new abstract mathematical structure that conforms to certain rules and that we can use to see whether we can somehow group the equations and their solutions and see whether they have solutions in radicals. Galois made a *group* of all the possible solutions or roots of an equation to work out whether such equations related to it can be solved. And that meant that you could use groups to study mathematical equations. This type of group is now called a *Galois group*.

A Galois group consists of the renumbering of all solutions of polynomials (equations). *Renumbering* is a certain procedure. Imagine you have an equation of a certain degree and you can find its roots (solutions). One of these roots is a positive radical, and the other one is a negative one. If we take that positive radical and use it in another expression, it transforms into a new equation. Do the same with the negative radical. If these two new equations are somehow symmetrical, this is related to the Galois group, and through such renumbering of solutions we can transform one relationship into another.

When we use Galois theory to study equations, their solutions and their coefficients, we form series of what we call now *fields of numbers*, and through these we can find, by quite a complicated method, a Galois group. The result will give us an answer to the question of whether an equation has a solution in radicals. This sounds rather complicated. But the work of Galois enables us to know in advance whether an equation *can* be solved – yet another short-cut, of which mathematicians are so fond! The study of groups eventually became incredibly important not only in algebra but in geometry too, in the study of symmetries.

Neither Abel nor Galois were recognised during their all too short lives, but they have been celebrated ever since. We now have Galois theory, the Abel–Ruffini theorem, Abelian groups, and for more worldly ways of expressing gratitude to mathematicians, the Abel Prize. Modelled on the Nobel Prize (which does not include one for mathematics), the Abel Prize is awarded by the King of Norway every year to a deserving mathematician who came up with a new discovery. Its value is the not inconsiderable 7.5 million Norwegian kroner (around £500,000, or $700,000) – riches that the prize's namesake could only have dreamed of.

CHAPTER 25

A Logical Wonderland

With all this talk of imaginary numbers, hyperbolic space, parallel lines that meet or may not exist at all, perhaps you're feeling a bit like Alice having tumbled down the rabbit hole into a mathematical Wonderland. On first glance, nothing seems to make much sense.

By the mid-nineteenth century, when Lewis Carroll (aka Charles Lutwidge Dodgson, 1832–98) was writing *Alice's Adventures in Wonderland*, mathematics was undergoing a huge transition. The way mathematics had been done since the third century BCE was slowly being replaced. Euclid was being questioned by the inventions of Bolyai and Lobachevsky. The study of equations could be supplemented or even replaced by the theories of Galois and Abel.

Although now known primarily for his children's stories, Dodgson was an exceptionally able mathematician, and remained at Christ Church, Oxford, after studying there. It's safe to say, though, that he was a bit of a traditionalist. He didn't much care for the new mathematical theories which were then starting to take hold. Some of that discomfort made its way into his Alice stories. Questions that have no answers, rules that have no basis, things transforming into other things – it all seems intended to playfully ridicule the new-fangled mathematics that was taking the field far away from a world where two and two make four.

Dodgson, who wrote so brilliantly about Alice's illogical world, was very interested in logic. He certainly knew of the works of George Boole (1815–64), who approached logic and how it might be applied to mathematics in a brand new way.

Boole was a man of great intellect and confidence, even though he was mostly self-taught. From relatively early on in his life he had to financially support his parents, establishing his own school when he was just nineteen and then running other schools in the north of England. All the while he was thinking and developing his mathematics, and though he couldn't take up a place at university because of his responsibilities, he did correspond with eminent mathematicians and gained a reputation as an outstanding mathematical scholar. Eventually he became the first professor of mathematics at the newly established university in Cork in 1849, teaching there for the rest of his life.

By then he had published some works on his thinking about the new relationship he could see between logic and algebra, which he had been developing since 1838. Logic is a branch of philosophy dating back to antiquity; Aristotle's *Organon* consisted of six works on logical analysis and dialectic. Logic deals with reasoning and is sometimes called also a science of deduction. Dialectic is about investigating, through arguments, the truths of opinions.

What Boole wanted to do was to reduce logic to a simple algebra, bringing it properly into the mathematical sphere. By connecting algebra and logic, Boole was going to bring back some of that certainty that Euclidean geometry had provided to millions of people learning mathematics for more than twenty centuries. Now that it wasn't certain any longer that Euclid's was the only system of unshakeable mathematical truths, Boole was trying to find another method by which we could be absolutely certain when something is true or not (as the case may be). What a stroke of genius, to connect logic with algebra! Especially from someone who had barely had any formal mathematics education.

Boole firstly introduced *binary logic*. This binary refers to two values – true, and not-true (or false). Everything, Boole thought, should be either true or not. True would be represented by 1 and false by 0. Next, he drew an analogy between algebraic symbols and those that represent logical ones. What he created became known as *Boolean algebra*. It is different from elementary algebra because the variables are those truth values – true and false only. In our ordinary, elementary algebra, the variables are represented by letters, but their values are numbers which you find by solving algebraic equations. But in Boolean algebra, what you are searching for is the value of variable that can only be true or false. In elementary algebra, we have symbols such as +, −, ×, ÷, =. In Boolean algebra, we have operations of conjunction (*and*) represented by ∧, disjunction (*or*) represented by ∨, and negation (*not*), represented by ¬. But the variables can only ever be true or false. Boolean algebra relies on this binary system of values.

Boole wrote a book on this, *The Laws of Thought* (1854), but he could not claim to have invented the binary system altogether. Long before Boole there had been developments that introduced the possibility of a number system which had only the values of 0 and 1. The binary number system in Europe went as far back as Leibniz, who wrote a treatise on the arithmetic of binary numbers in 1705. But what Boole did was to liberate the method from numbers. After Boole assigned the two values to *true* and *false* you could do quite a lot more with it. Algebra of huge complexity could and was developed from the basic operations he mapped out. It took some time to do that, but Boole recognised it in the first place, which was crucial.

With the two values of 0 and 1 in Boolean algebra, its correspondence with the base-2 number system was very obvious. But, you might ask, how could you possibly connect 1s and 0s with our decimal number system? What about the numbers 2 to 9? Only having 0 and 1 seems rather restrictive. It is, however, possible to 'translate' numbers from decimal to binary. Rather than the decimal system, which is base 10, the binary system is base 2, as we just have two choices (0 or 1). To translate any number in the decimal system into binary, we break it down into its different powers of 2, and write 1 when it does have a particular power of 2, and 0 when it doesn't. The numbers look quite different, but they are equivalent: for example, 23 in base 2 comes out as 10111.

A direct legacy of Boole's investigations into the mathematical representation of logic is the widespread use of binary numbers in computer information processing today. The birth of computers came not long after Boole, when another British mathematician dreamed of making modern and very capable calculating machines – what he called the Analytical Engine and the Difference Engine. This was Charles Babbage (1791–1871), who some now call the 'father of computers'. His design for the Difference Engine, which was a mechanical calculator, was the simpler of the two. The Analytical Engine, meanwhile, was more along the lines of our understanding of a computer – it could do various logical operations, and had a form of integrated memory. Babbage designed these two engines, but didn't succeed in completing the construction of them. His partner in this pursuit was Ada Lovelace (1815–52), a mathematician, writer, aristocratic socialite and only daughter of the poet Lord Byron. She greatly promoted Babbage's work and came up with new ideas about calculating machines – she thought such engines could be used not only to perform numerical calculations but deal with other quantities, even music or letter forms. Both of Babbage's engines were mechanical devices and neither used the binary system, but nevertheless they sowed the seeds of something new. Around a century later, Claude Shannon (1916–2001), an American mathematics student, linked the two. In his 1937 thesis, *A Symbolic Analysis of Relay and Switching Circuits*, he used Boolean algebra and its logical principles and applied it to electronic circuits (where 0 would mean 'no' and 1 'yes'). From the early ideas of Babbage and Lovelace, Shannon and the many other mathematicians and computer scientists following him would eventually bring Boolean algebra into most of the homes and offices on our planet – via personal computers.

Back in the nineteenth century, developing other representations, particularly diagrams, became all the rage – Dodgson contributed some diagrams that helped with logic during his own research. But it was John Venn (1834–1923) who gave his name to diagrams of a particular sort that we use so often today.

Venn was younger than Boole, and studied mathematics at Cambridge. He became so disgusted (in his own words) by the way the university examined mathematics and taught students only to pass those exams that he sold his mathematics books and said he would never come back to the subject. He became a priest in 1859 and served in two towns in England. But he returned to Cambridge to teach moral sciences, and there came across Boolean algebra in the 1860s and 1870s. He was so enchanted by this new mathematical invention that he broke his own promise! He wanted to work further on this and improve it if he could. In 1866 he wrote *The Logic of Chance*, therein using logic to underline probability theory.

A Venn diagram, as they are now called, is a method you can use to represent many operations between sets of objects using Boolean operations. The beauty of this representation is that you don't have to have numbers as elements of a set, although you can. Using a Venn diagram, you can represent what kind of relationship there is between two sets, A and B; in particular, if they share elements, make them overlap. Boolean operations could then show, for instance, that $A \wedge B$ for certain elements in the set (because they can be found in both sets, at their intersection where they overlap), and $A \vee B$ for other elements that are *either* in one *or* the other set. Venn diagrams can also map out more complicated processes. By breaking down two numbers into their prime factors and making a diagram from them, you can see where they overlap, and that gives you the two numbers' highest common factor (HCF). Multiplying the HCF by the remaining prime factors gives you the lowest common multiple of the two numbers.

Venn's diagrams and Boolean algebra became very important in the new mathematics to come. The focus of mathematics changed from looking directly at mathematical objects, such as parallel lines and higher-degree equations, to understanding their place and relationships in a much wider framework. This new mathematics would eventually give rise to entirely new fields of science. Being able to develop algebra which was based on two values of *true* and *false* only, when applied later to electrical circuits, meant that you could have two states that could be translated to *yes* and *no* – there *is* electrical current or there *isn't* one. This wasn't thought of until 1937; electricity hadn't even been discovered when Boole came up with his invention. But in time, Boolean algebra made the storing and manipulating of information by digital computers possible. So I have him to thank for being able to write this book for you.

CHAPTER 26

The Quiet Birth of Chaos

'A scientist worthy of his name, above all a mathematician, experiences in his work the same impression as an artist; his pleasure is as great and of the same nature.' This is a famous quote by one of the famous mathematicians of the end of the nineteenth century, Henri Poincaré (1854–1912).

Is it true? Is a mathematician's work like that of an artist, and if not, how does it differ? Many people would think the two have absolutely nothing in common. Where is the aesthetic pleasure in maths of all things? But most mathematicians believe that art and mathematics are very alike. They are both creative activities, requiring inspiration and understanding that in turn makes their work understandable and useful to others. We have already used words like 'beautiful' to describe some of the proofs that mathematicians have come up with. And there *is* some kind of artistic pleasure involved here – in making and appreciating mathematics that is elegant, insightful, orderly, surprising, deep, somehow inevitable, or whatever it is. But sometimes you have to 'do the math' to be able to see it.

One of the many things that Poincaré was interested in was the mathematics of celestial motion – planets and stars in space. He completed his doctorate in Paris in 1879, and eight years later participated in a competition to celebrate the sixtieth birthday of the King of Sweden and Norway. The question was formulated by the Swedish mathematician Magnus Gustaf 'Gösta' Mittag-Leffler (1846–1927), with a prize to anyone who could come up with a solution. The question concerned what was then known as the n-body problem. How could one describe the forces acting between any number (n) of celestial bodies mathematically?

Now, as we have seen, Kepler and Newton had already made great strides in understanding the movements of those celestial bodies in the solar system. Their masses, speed and direction of travel were known, their future motion described by Kepler's laws of planetary motion and Newton's laws of universal gravitation. But the gravitation *between* the celestial objects, attracting each other this way and that, was precisely the problem. The issue was the *interactive* gravitational forces. While the forces themselves did conform to Newton's laws of motion and universal gravitation, the interaction of those forces in the maelstrom of space was difficult to predict. Finding a general theory for any number of celestial bodies seemed nigh-on impossible, and that was what the question, posed on behalf of the Swedish king, was all about.

Henri Poincaré's entry didn't precisely tackle the general n-body problem, but rather the three-body problem. Before the King of Sweden's birthday challenge, other mathematicians had already made some advances on this question, including Newton, via Jacob and Daniel Bernoulli, Euler, Lagrange and more. But Poincaré's submission involved some ingenious and inventive mathematics and introduced new techniques that could be expanded upon. It was enough to win him the prize. He started preparing the submission for publication in *Acta Mathematica* on the king's birthday, but realised that his paper had serious errors in it. He withdrew it, revised it, and finally published it in 1890, continuing to work on the problem for the next ten years. Poincaré stated that there is no general solution to the three-body problem that could be expressed in a finite form.

During his revisions of his original entry, Poincaré also realised that the three-body state was greatly affected by initial conditions: very slight differences here could lead to drastic differences later on. Even tiny variations in the mass, location, speed or the direction of motion of the celestial bodies would accumulate over time, and eventually lead to unpredictable or chaotic behaviour. Poincaré realised the importance of this discovery. He suggested that this behaviour is probably widespread in nature, including meteorology, the science of weather.

At the time this issue of initial conditions was recognised and researched further, but the computational power of people is nothing compared to that of actual computers. It took more than half a century for another scientist, mathematician and meteorologist Edward Norton Lorenz (1917–2008), to be able to compute his weather models. He described how a small change in a system can eventually result in enormous, unpredicted differences elsewhere as the *butterfly effect*. It is so called after a metaphor Lorenz used to describe the behaviour: that the slight changes made to the movement of air by a butterfly flapping its wings could eventually result in a tornado on the other side of the planet. This didn't, and doesn't, happen in reality, but it does point to the idea that drastic changes can occur after the tiniest initial changes or disruptions.

The butterfly effect is a metaphor that describes an underlying principle of *chaos* and *chaos theory*, a whole new branch of mathematics that deals with complex systems, just like the *n*-body system, whose behaviour is very sensitive to its initial conditions. This has now become an area of great interest to mathematics, and it all started with Poincaré. You could say that his mistake during his initial work on the three-body problem – so tiny in the scope of the whole history of mathematics – caused a butterfly effect that led to the invention of chaos theory.

In the same year that Poincaré was born, another mathematician was already exploring the mathematics of space. The German mathematician Bernhard Riemann (1826–66) studied non-Euclidean geometry but extended it into higher dimensions. Multidimensional space is also possible, said Riemann – not only that, but it is essential for mathematicians to think about this!

He defined a field of geometry we now call Riemannian geometry. It is concerned with the study of things called *manifolds* (now named after him, Riemannian manifolds). A manifold is a topological space that, locally, might look like Euclidean space, but which isn't really Euclidean in general. A sphere is a manifold – the angles of a triangle drawn on its surface add up to more than 180°, although a small triangle would approach a shape where Euclidean geometry is a good approximation. Riemann explored smooth manifolds through his new geometry (a version of differential geometry) based on the idea of curves, surfaces and their relation to topology. Let's recall that in topological space, the relationships between things matter, but not their magnitudes: you would consider closeness, but not distance. So in a Riemannian manifold, for instance, you could measure its curvature and find how it differed from flatness at that particular local area.

We could easily get very lost in Riemann's geometry! So let's content ourselves that Poincaré knew it well, but wanted to look at something that wasn't either Euclidean or Riemannian geometry. Riemann had described a model of space as a disc, but didn't pursue it much further. Poincaré now took that up and made it widely known.

How can space be a disc? Well, this is a model, and a mathematical model at that. Poincaré didn't mean it represents actual space; it has a different geometry to that of our experience. But he did want to see where such a model would lead and what new mathematical laws it would reveal. He wanted to push the boundaries of our imagination in many ways. His *Poincaré disc model*, a whole new model of hyperbolic geometry, very much does that. Let's look at how it works.

Poincaré's disc is a circle that encompasses hyperbolic space. The boundary of that space, at the edges of the disc, is infinitely far away from the middle. Shapes that appear towards the disc's edge get smaller and smaller and eventually will become just dots on the disc's circumference as it reaches infinity. We can picture it, but bear in mind this disc is infinitely large, so it wouldn't look exactly like this.

This model opens up all kinds of things that make the geometry consistent within its system, but look a little strange to us if we only think of space as Euclidean. One thing is what it does to straight lincs. For us, a straight line is the shortest distance between two points. In Poincaré disc space, though, straight lines are at a right angle to the boundary, the edge of the disc. The diameter of the disc satisfies this, and looks reassuringly straight, as we would expect. But straight lines here can also be circular arcs that are orthogonal (at right angles) to the edge of the disc.

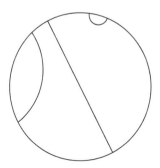

What happens to triangles? In Euclidean space, the sum of angles in a triangle is equal to exactly two right angles, but in Poincaré disc space it's less than two right angles. And, because we are talking topologically here, in this model of space, all similar triangles are congruent (exactly the same), because there is no measurement of length as such. Whereas in Euclidean geometry triangles that are congruent have to have all three angles and all three sides exactly the same, in this *disc* model, if the triangles have the same angles, they will be of the same size, and hence equal to each other.

We've already said that topology is about the relationships between space and not about measurement. Shapes are also *seen* differently in topology; things that look very different can be compared. In topology, you can stretch and squash objects, but no cutting, sticking, or making or closing holes is allowed. In this branch of mathematics, you can imagine a sphere morphing into a cube or cuboid – in topology they would be *equal*. The geometrical properties of a figure are not affected by changes like making or closing a hole in them. No cutting, sticking or making holes is allowed, but you can squash and stretch. So for a topologist, a coffee mug (with a handle) and a ring doughnut are one and the same. What's important is that both objects have just one hole and no more: there's one hole in the mug (formed by the handle) and one hole in the doughnut (as long as it's not the sort filled with jam!). There's a popular joke about a topologist who didn't know whether to dunk a doughnut in his mug or his mug in his doughnut . . .

Perhaps you would think that inventing (mathematical) chaos and a new model of space would be enough for one lifetime. But Poincaré is best known today for something we now call the *Poincaré conjecture*, which he stated in 1904.

It says that if you have an object without holes, and it is relatively small – that is, it doesn't go infinitely in any direction – it is (topologically) a sphere. No arguments so far: any lump of clay can be moulded into a sphere and squashed and squeezed into umpteen other shapes that are topologically still a sphere. But what Poincaré said next was the difficult bit: he stated that this holds in *any number of dimensions.* This quickly gained a reputation as being particularly tricky to tackle, not only difficult, but making mistakes in proofs very hard to spot. In the latter half of the twentieth century people managed to prove that this holds for five dimensions and above, but no one had managed to prove it for the fourth dimension. (Imagining a fourth dimension can be a little hurtful for your brain, but we'll try to do this in Chapter 28.)

The Poincaré conjecture was included as one of seven questions the Clay Mathematical Institute set as Millennium Problems in the year 2000, offering $1 million to anyone who solved one of them. Just three years later, an argument sufficient for proving the conjecture, based on new and revolutionary understanding of related mathematical problems, was posted online by Russian mathematician Grigori Perelman (1966–). In the years to follow it was discussed further and mathematicians interrogated his somewhat terse and very complicated solution and found it to hold. Perelman had not only won a Millennium Prize of $1 million, but was also awarded a Fields Medal, often termed 'the Nobel Prize of Mathematics', which since 1950 has been awarded every four years to a mathematician under the age of forty who has made a major contribution to mathematics. It is one of the highest honours a mathematician can receive. But what happened next shocked the global mathematical community. Perelman declined both awards, the Millennium Prize and the Fields Medal. He didn't like that the judges hadn't recognised the other mathematicians who had contributed to the solution by discovering things he relied on before him. He considered the Fields Medal to be, he said, completely irrelevant. The greatest prize for him was that he proved this conjecture. No other recognition was needed.

CHAPTER 27

Sizes of Infinities

Mathematicians' thinking about infinity has its beginnings, curiously, in very tiny quantities. It starts with limits, which are at the very core of the nature of calculus. When we try to find a derivative of a function (Chapter 20), we are looking at infinitely small values, so finding those points where something tends to a particular value (a limit) is crucial. And mathematicians like to be extra rigorous when doing thinking like this, so very small quantities matter a very great deal.

Over the centuries, mathematicians had reformulated the definition of the nature of limits while working with ever more complex calculus. For example, ever since antiquity, mathematicians were using the idea to work out to which value a repeated calculation would approximate. Archimedes used it to give an approximation of π to be somewhere between $3\frac{1}{7}$ and $3\frac{10}{71}$ by calculating the ratio between a side of a polygon and its long diagonal, with an ever increasing number of sides (therefore approaching circle and modelling the ratio between diameter and circumference of a circle). We have come across limits earlier – it is a number that a number sequence or series tends towards but never quite reaches. Some of these series are quite simple, like the series of reciprocals of the powers of 2 (Chapters 11 and 21): $\frac{1}{2}+\frac{1}{4}+\frac{1}{8}+\frac{1}{16}+ \ldots = 1$. In this example the limit is a very clear and finite number – it is just equal to 1.

But even here, there is a potential problem. Where, you may ask yourself, do these fractions stop? You could go for ever, writing fractions that represent smaller and smaller values. What that means is that you will eventually have an infinitely small value, a fraction with an infinitely large denominator. When $n \rightarrow \infty$ – that is, when n tends towards infinity – then $\frac{1}{n}$ will become an infinitely small value.

Then there was the question of infinity at the opposite end – the infinitely large numbers. Is this infinity the same as the infinity we're talking about with infinitely small values? How can we tell? This might prompt a whole string of related questions. If there are these two infinities, are there any others? Is there an infinity in the depths of negative numbers? How many infinities are there, after all? The simplest answer is – infinitely many. But there is a mathematically rigorous answer to this too, which became incredibly important towards the final decades of the nineteenth century.

Georg Cantor (1845–1918) was the first mathematician to fully address the abstract concept of infinity. He was born in St Petersburg but moved as a child with his family to Germany, where he remained until the end of his life. He was fascinated by infinity from a young age, and from around the age of thirty began publishing his mathematical ideas about it. This was the start of modern set theory.

Cantor wrote an important paper in 1874 to introduce his theory of sets. A set is any collection of separate individual objects. We can compare sets and say that they are equivalent, or that they have the same size, if we can match them in a one-to-one correspondence. We have to order the sets in some way in order to compare them with other sets. When we compare two sets, each element in the first set should have an equivalent element in the second set. Now we can look at the sets in terms of size. There are finite and infinite sets. Say we have a set of four pencils. This is a finite set, and it has a cardinality of 4. If I look at a set of four children, I can assign each child exactly one pencil, and that set has also a cardinality of 4. *Cardinality* is the way of comparing sizes of sets, and the sets with the same cardinality we say are equivalent. Why not just say these sets both have a size of 4? Because there is a cardinality of infinite sets too! And infinity doesn't have size. Or does it?

Now let's look at infinite sets. Cantor wasn't the only one to think about infinities; his friend Richard Dedekind (1831–1916), for example, was also working on this. But Cantor was the first to try to find the sizes of infinite sets. Was this not all a bit silly? Some of Cantor's contemporaries thought so. Because infinite sets are infinite, they would go on for ever. If we look at the sets of natural numbers, they start from 1 (sometimes 0 depending on whatever convention is being followed). The set of natural numbers we denote with $\mathbb{N}$ and we know it goes on for ever. We would write this as $\{1, 2, 3, 4, 5, 6, \ldots\}$. But we can look at another infinite set, for example a set of square numbers $\{1, 4, 9, 16, 25, 36, \ldots\}$. Are the infinities of these two sets equal or is the infinity of natural numbers larger than the infinity of square numbers?

We said that to work out whether some sets have the same size then their elements need to have a one-to-one correspondence. Let's try to do that with these two sets. The natural number corresponds to its square (1 goes 1, 2 with 4, 3 with 9, and so on).

1	2	3	4	5	6 ...
1	4	9	16	25	36 ...

We have linked the two sets' elements, but does that mean the sets are of the same cardinality? There are so many more numbers *in between* the square numbers, so perhaps the infinite set of natural numbers *is* larger than the infinite set of squares. But how could we possibly know by how much?

Cantor came up with a little principle here that proved to be very useful in thinking like this. Let's call the sets we can list or enumerate *infinitely enumerable*. If we use the set of natural numbers as a kind of reference set, then we can say that if we can establish one-to-one correspondence with it, that second set (like our square numbers) will also be infinitely enumerable. Because we can establish this correspondence, these two sets, although infinitely large, are of the same size, or rather, their infinities are of the same size.

But Cantor, and some of his friends and colleagues, including Richard Dedekind, couldn't leave it there. They wondered whether there were any infinite sets that are of a different size of infinity than the set of natural numbers. Are there in fact different sizes of infinity in general, and if so, how could mathematics define and describe them?

First we need to mention some of these sets of numbers so we know what they are like and which types of infinite numbers they contain:

$\mathbb{N}$ is the set of natural numbers: $\{0, 1, 2, 3, \ldots\}$
$\mathbb{Z}$ is the set of integers: $\{\ldots, -3, -2, -1, 0, 1, 2, 3, \ldots\}$
$\mathbb{Q}$ is the set of rational numbers, which are all numbers you can express as fractions of two whole numbers
$\mathbb{R}$ is the set of real numbers, all the numbers on the number line; it contains all the previously mentioned sets, and a few more . . .

There are algebraic numbers, sometimes denoted by $\mathbb{A}$. Algebraic numbers are solutions to algebraic equations. All rational numbers, integers and natural numbers are algebraic, and so are a lot of irrational numbers. Two numbers that are non-algebraic are π and e. You can't have an equation to give you a precise value of either π or e. We have formulae for π and e which define what they are equal to, but these are not equations. In other word, they are *transcendental* numbers – they transcend algebraic numbers.

All of these are real numbers. They will all have a place on the number line.

But what sizes are these infinitely large sets of different types of numbers? Let's look at the integers first. Surely because they go in both directions (positive and negative) into infinity then their infinity is larger than the infinity of natural numbers? However, you can still simply list the integers in such a way that you establish a one-to-one relationship with natural numbers:

0	1	−1	2	−2	3	−3	4	−4 . . .
|	|	|	|	|	|	|	|	|
1	2	3	4	5	6	7	8	9 . . .

So the set of integers is also an enumerably infinite set.

Surely the set of rational numbers is not like that! How would you even begin to list all the possible fractions? Cantor came up with a way of listing even these elements. It looks a little strange to see it for the first time, but it is a very simple method. He starts with $\frac{1}{2}$ and $\frac{2}{1}$, where the numerators and denominators swap, but in both cases the numbers add up to 3. Continue in the same way but increase the sum, so the next fractions would be $\frac{1}{3}, \frac{2}{2}, \frac{3}{1}$, with the sums of numerators and denominators adding up to 4. And so on. In this way, because you can enumerate all the natural numbers, you will eventually enumerate all the possible rational numbers. And you can establish a one-to-one correspondence, which means that the cardinality, or size, of the set of rational numbers is the same as the set of natural numbers. Cantor gave this cardinality a name, *aleph-null*, written as $\aleph_0$. Cantor chose the first letter of the Hebrew alphabet, possibly harking back to his Jewish origins.

But what about the set $\mathbb{R}$ of real numbers? Some believe that this question bothered Cantor from the very start and that this stimulated the birth of set theory in the first place. Imagine a number line, with 0 in the middle and numbers stretching in both directions into infinity. Looking between the labels 1 and 2, you could imagine where 1.5 would lie. But keep looking, zooming in closer and closer – so many more numbers would appear between those labelled. You would see more and more numbers and some of those would be numbers with decimal places that never end. You'd discover an infinity just between 1 and 2. But then, look, there's also an infinity between 1 and 1.5 . . .! And so on and so on until, you guessed it, infinity.

This is where the problem lies – you can't stop zooming in, there's always another number there. With real numbers the one-to-one correspondence with the set of natural numbers couldn't be worked out. Cantor did try. He attempted a proof by starting from two numbers on the number line, say a and b. Then, as we did, he looked between them, finding more and more numbers between a and b. Is there an end to such 'zooming' in? Will we ever get to the last number? No, said Cantor. Whatever last numbers you find, say a_n and b_n, you can always find their average $\frac{a_n + b_n}{2}$. That means that this infinity is *non-enumerable*. It is not comparable to $\aleph_0$, the size of infinity defined by natural numbers.

Cantor's friend Richard Dedekind came to look at this. He developed the concept that we now call *Dedekind cuts*. This is a method of making a real number (any number on the number line) from rational numbers (expressed as a fraction of whole numbers).

Let's say we have some real number that we denote with x. All the rational numbers smaller than it (so before it on the number line) we can say will belong to the set A_x, and all the rational numbers larger than it (after x on the number line) to the set B_x. Dedekind *cut* the number line in two parts, and put these into two sets. The real number x in a sense 'fills the gap' between the two sets. We can't know what the precise value of x is, and so we can't know the exact values of the numbers either side of it (they are infinitely near!). Dedekind cuts are important as they literally bridge the gap – they provide us with the only method (so far discovered) of dealing with these 'holes' in the number line, as we can nevertheless talk about the real numbers by defining them by rational numbers (those on either side of them on the number line).

Once Cantor established that there is no one-to-one correspondence between natural and real numbers, he showed that there is more than one level of infinity in mathematics. This is sometimes called the *continuum hypothesis*. It states that the size of the infinity of real numbers is greater than that of the natural numbers. And, further, that there is no third infinite set whose infinity (cardinality) is between these two.

We have entered quite an abstract domain now, which is why modern mathematics is not often taught at school. But from all this talk of infinities, there is one conclusion: Cantor proved that given an infinite set, there is always a set whose cardinality is larger. This means that there is no maximum size of infinity. Which means there are infinitely many infinities – after the largest you find, there will always be another that follows.

CHAPTER 28

The Unfolding of Many Dimensions

Do you ever feel stuck in this world and wish you could escape into unexplored dimensions for a while, just for a little adventure? Lots of science fiction stories explore that – travelling through time or visiting parallel universes. Perhaps some of the mathematical ideas we've already encountered have made our three-dimensional world with all its homogenous space feel, so to say, a little flat.

We don't have to rely on stories to experience different dimensions. Mathematics can offer it too. In fact, thinking about higher or more dimensions has been something that mathematicians have grappled with for many centuries. Let's time-travel some earlier mathematicians into our chapter now. From the seventeenth century, English mathematician John Wallis (1616–1703) said that even to imagine a dimension which is not one of the three usual ones – length, breadth and thickness – is a 'fansie', and that an object of four dimensions would in effect be a 'monster in nature, and less possible than a Chimaera or Centaure'.

Indeed, this was the prevalent view in mathematics for centuries. Then, in France, just before and at the start of the French Revolution, Jean le Rond d'Alembert (1717–83) and more fully Joseph-Louis Lagrange (1736–1813) put forward the idea that *time* could be seen as the fourth dimension. If you think about it, any object which exists in space is somehow defined in time too, as time and change are interdependent. But for Lagrange, this fourth dimension was very much centred on the mechanics of geometry.

Later mathematicians were very interested in those monsters described by Wallis, though, and developed the concept of higher dimensions when looking at objects in space. The start was the investigation of three-dimensional shapes, or polyhedra. A polyhedron is a three-dimensional shape that has flat faces, straight edges and sharp vertices. A *convex* polyhedron is such that, if you were to draw a line segment between any two points on its surfaces, the segment would *only* go through its interior. This was one finding of the prolific and brilliant Leonhard Euler (Chapter 21). Not all polyhedra are convex – recall the star polyhedra, such as the small or great stellated dodecahedron (Chapter 15). Some of the segments that would link two points on their surfaces would definitely go outside of them. These are *concave* polyhedra.

In his 1758 paper *Elementa dotrinae solidorum* (*Elements of the Doctrine of Solids*), Euler described one of the most important characteristics of convex polyhedra. He worked out that the number of vertices, minus the number of edges, plus the number of faces, would always give you 2. Or, as a formula, $v - e + f = 2$. A cube is a convex polyhedron. It has eight vertices, twelve edges and six faces, and $8 - 12 + 6 = 2$. At first sight this might not seem all that relevant to anything, aside from the finding that there was some pattern in the way convex polyhedra are made. But it was the crucial jumping-off point for another Swiss mathematician, Ludwig Schläfli (1814–95), in his work on describing polyhedra and his investigations into higher dimensions.

Schläfli's masterpiece, *Theorie der vielfachen Kontinuität* (*Theory of Multiple Continuity*, 1901), was his attempt to found a new multi-dimensional geometry (in a different realm from three-dimensional space with its three-fold 'continuity'). Sadly it was not published until after his death, and only then was its importance recognised. In this work he introduced *polyschemes* (what we call *polytopes*), which were higher-dimensional regular polyhedra. Just as Euclid proved there are five regular solids in three dimensions, Schläfli proved there are six regular solids in four dimensions, and three when we're talking about the fifth dimension and higher! He also showed that Euler's formula extends to these higher dimensions: that the minus should be followed by plus, and so on, interchangeably. If this pattern continues as far as one would wish to go with it, the result will still always be 2.

The study of the fourth dimension was in vogue in nineteenth-century mathematical circles. Quite a few mathematicians from this period in some way touched or directly influenced the developments in working out mathematical laws for higher dimensions. Arthur Cayley, who we already met in Chapter 23, wrote about this emerging field in mathematics in *Chapters in the Analytic Theory of n-Dimensions* in 1843. Various German mathematicians, such as Hermann Grassmann (1809–77) and Julius Plücker (1801–68) worked on this too. Bernhard Riemann gave a now-famous lecture on curved space in 1854, in which he discussed how an *n*-dimensional space can be defined using what we now call *Riemann space* (as we saw in Chapter 26, mathematically defined spaces that locally are Euclidean but in general may be non-Euclidean).

Towards the end of the nineteenth century, two very different methods of enquiry intersected in this particular topic. On the one hand there were scientists who investigated phenomena linked to electricity, magnetism and astrophysics; on the other, spiritualists who developed beliefs on humans' ability to transcend this world and communicate with people from another. There were quite a few scientists among this latter group too, but its most prominent members, among them Arthur Conan Doyle and Charles Dickens, were from the art world. What does such a movement interested in seemingly spurious connections between this and the 'other world' have to do with mathematics? Quite a lot as it turns out.

Among the London circle of spiritualists were some remarkable mathematical men and women. Mathematician Charles Howard Hinton (1853–1907) was friends with Edwin Abbott Abbott, author of the great mathematical novella *Flatland* (1884), which imagines a journey from a two-dimensional world through different dimensions. Hinton himself would take such a multidimensional voyage in his research. His father's secretary was Mary Everest Boole (1832–1916), by then the widow of the famous George Boole (Chapter 25). Mary was herself a self-taught mathematician and mathematics teacher-writer, and she encouraged the teaching of mathematics to all children through play. She certainly instilled a love of the subject in one of her five daughters, Alicia, later Boole Stott (1860–1940). She had an amazing talent for mathematics. Alicia was only four when her father died, and never had any formal training in the subject, but growing up in her mother's care would have made up for the loss and restrictions on women accessing education at the time. As a child she was also taught mathematics by Charles Hinton, particularly geometry related to the fourth dimension.

Hinton studied mathematics at Balliol College, Oxford, finishing his BA in 1877, after which he went to teach at Uppingham School in Rutland. His interest in the fourth dimension was broad. He connected it with the interests of his spiritualist friends in accessing otherworldly consciousnesses. He wrote a few science fiction stories too, so as to encourage thinking about the translation from three dimensions to four. And he brought extraordinary insight to the further development of mathematics of the fourth dimension. In his most famous work, *The Fourth Dimension* (1904), he also instructed his readers how to visualise this dimension by 'casting out the self'. This is difficult to explain, and the difficulty is not only yours and mine but everyone else's. Conceiving of a dimension outside the three that we can see and inhabit is notoriously difficult. The only person who seemed to grasp this procedure and geometry was Alicia. She too could visualise the fourth dimension.

Scandal then befell the Hinton family: Charles was convicted of bigamy in 1886. He had married Mary, Alicia's older sister, in 1880, and had four children; it was now revealed that under a false identity he had wed again and had twins. The London circles he moved in were outraged. After Hinton was released he found it hard to find work. He took his (first) family to the newly opened Japan, then immigrated to the US, where Hinton worked at the universities of Princeton and Minnesota, and at the US Naval Observatory in Washington DC. Although he found some success in touting an invention of his, a gunpowder-operated baseball pitching machine, his mathematical career never really recovered. The story goes that, during one lecture on the fourth dimension, he said he would step into it, upon which he duly fell down dead.

Unlike her sister Mary, Alicia found a partner who seems to have supported her in all her endeavours. She married Walter Stott, an actuary, in 1890, and soon had two children. A couple of years later, in 1894, it's reported that it was Walter who got in touch with Pieter Schoute (1846–1913), a Dutch mathematician known for his work on regular polytopes and Euclidean geometry. Schoute even came to England to work with Alicia and they discussed the fourth dimension, in which Alicia was by now an expert in visualising objects. They collaborated together for almost twenty years, each bringing their unique strengths to the research, spending the university summer holidays together deep in discussion and publishing papers on topics related to the fourth dimension. Schoute's university in Groningen bestowed an honorary doctorate upon Alicia Boole Stott in 1914, a year after Schoute died, in recognition of her contribution.

Following her friend's death, Boole Stott took a break from mathematical research. Around fifteen years later, her nephew Geoffrey Ingram Taylor (1886–1975) brought a mathematician friend from Cambridge, Harold Scott MacDonald Coxeter (1907–2003), to meet her. Despite the almost fifty-year age gap between them, Alicia and Harold hit it off famously. They spoke and wrote to each other on four-dimensional geometry for the rest of Alicia's life.

For years, Alicia Boole Stott had made sets of cardboard models representing three-dimensional images or 'traces' of four-dimensional objects. (Some of these can still be found in various universities in the UK and Holland.) How could one make three-dimensional models of four-dimensional objects? Consider looking at a cube which travels in a uniform motion towards a sheet of paper; at some point it would 'pass through' the paper, and in doing so leave a trace. That trace we could then imagine being permanently drawn on paper – that would be a two-dimensional trace of a three-dimensional object. In similar fashion, one can imagine a four-dimensional object passing through a three-dimensional space and leaving a trace in the form of a three-dimensional model. These are the types of models that Boole Stott made.

She was incredibly good at this art. She was not only able to imagine and craft these models, but in her mind's eye she was able to calculate all their characteristics. She introduced the term *polytope* to refer to uniform polyhedra of any dimension. (She came up with this name, not knowing of Schläfli's term polyscheme.) She made a profound influence on the young Harold Coxeter, who went on to become a famous mathematician in his own right.

Boole Stott's legacy has only recently been recognised. As she didn't hold a university position, her research and published works were scattered around different places. A paper roll containing her drawings of polyhedra was only unearthed at Groningen in 2001. But now it's clear that her incredible gift to visualise and model objects in the fourth dimension enabled others to see much further too.

The story of the fourth dimension doesn't end here. This was only the beginning. Now it's an accepted practice that, once you find a mathematical rule or theorem that is valid in two or three dimensions, you see whether it can be proven in the higher dimensions. Rarely are today's mathematicians happy with just the four dimensions, even.

Have we ever seen the fourth dimension in our world? Perhaps. In Paris, the enormous La Grande Arche de la Défense (originally La Grande Arche de la Fraternité) models the fourth-dimensional hypercube projected onto our three-dimensional space. This architectural masterpiece was intended, its architects said, to provide a window to the world. Certainly, after the research and revelations of Hinton and Boole Stott, mathematicians now had their eyes trained on the furthest vistas of the higher dimensions.

CHAPTER 29

Mathematicians of the World Unite

There were so many different geometries by the 1890s that one mathematician wanted to sort them out, and needed to enlist the help of the international community of mathematicians to do it. Let's just absorb that: in less than a century, mathematics had gone from having one single (Euclidean) geometry to such a panoply of geometries related to the new theory of groups that they had to be classified in order to be properly made sense of. That's quite astonishing.

The mathematician spearheading this was Felix Klein (1849–1925). He had a sense that geometries could in some way be classified in relation to the groups of symmetries they had. In the 1870s Klein had made significant advances connecting Euclidean and non-Euclidean geometry in such a way that the latter – at that point still a pretty controversial subject in the wider mathematical field – was put on an equal footing with the former. One fascinating physical object devised by Klein, and named after him, illustrates his period's turn from a classical understanding of space to a new one. Although we can make a version of the Klein bottle in three dimensions (and people have), it is actually a two-dimensional manifold. If you imagine tracing your finger across its surface, the outside becomes inside and vice versa as the surface loops back though itself. There is in fact no inside or outside of it!

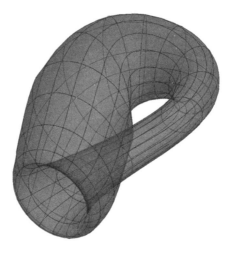

This is often mentioned in conjunction with a Möbius strip (named after the German mathematician August Möbius, 1790–1868), which is a two-dimensional equivalent, a non-orientable surface you can get when you attach the ends of a strip of paper with a half-twist.

Klein was an excellent mathematician himself, but also good at bringing mathematicians together, a little like Mersenne was in his time (Chapter 17), and Paul Erdős was in his (Chapter 31). When Klein came to the University of Göttingen in 1886 it was following a fairly serious physical and mental collapse. It marked a change in direction for him, and he poured his energies into making Göttingen into a world-leading research centre for mathematics. He succeeded in opening the doors to women, encouraged discussion groups and the development of a mathematical library, and under his management transformed *Mathematische Annalen* into one of the leading mathematical journals of the day. He brought in his friend David Hilbert (1862–1943) as a professor; the two men had already collaborated, and would continue to do so, the young man soon to prove himself to be one of the most influential mathematicians of the time.

While all this was going on, Klein had the idea for a programme that the international mathematical community should be working on together. Georg Cantor (Chapter 27) had been the first to press for regular meetings of mathematicians from different countries (to start off with, France and Germany), where new mathematical discoveries and developments in the field could be discussed, and Klein and others readily lent their support. The first International Congress of Mathematicians took place on the neutral ground of Zürich in Switzerland in 1897. This was a trial run of sorts, but went very well, with over 200 mathematicians from sixteen countries attending.

It was at the second congress, which took place in Paris in 1900, that David Hilbert gave what many mathematicians agree is the most famous lecture in the history of modern mathematics. On 8 August, his rousing speech, 'The Problems of Mathematics', described ten of twenty-three key unsolved problems, now called Hilbert's problems. Hilbert saw unanswered, even unanswerable, problems like this as the very opposite of problematical. These were the very lifeblood of mathematics; they gave the subject and those practising it vitality. 'We hear within ourselves the constant cry,' he said. 'There is the problem, seek the solution. You can find it through pure thought.'

Hilbert's problems lay down a challenge to mathematicians and set the research agenda for twentieth-century mathematics. Every time a solution to one of the problems has been found it's been celebrated as a major mathematical event. (Ten currently remain unresolved.) Considering the greatest mathematicians of the world have set their minds to these, the problems are really very hard, but let's still try and get a feel of some of them.

The first problem on the list was the *continuum hypothesis*. This states that there can be no set whose cardinality (size) is strictly between the cardinality of the set of natural numbers and the set of real numbers. We looked at some of the mathematical wrangling over the sizes of infinities in Chapter 27, but this hypothesis was later shown to be not provable using the standard axioms in mathematics, and a bit later than that, not disprovable either! That meant it could be taken off the list. But it had important ramifications for set theory, as we'll see in the next chapter.

Hilbert's third problem was actually the first to be resolved. This asked a question about *equidecomposability*. This term means that some objects should be equally composable and decomposable. Say we have two different polygons of any shape (but with three vertices or more), and which have the same area. We know that we can *decompose* them by cutting one of them into a number of triangles, then cutting and arranging the triangles to get rectangles, then cutting and rearranging again to get the shape of the second polygon. You may have to cut your shape into many tiny pieces, but you will get there eventually.

Hilbert's question was whether this could be done also in three dimensions. Does this equidecomposability hold for three-dimensional polyhedra? If you have two solids of the same volume, say a tetrahedron and a cube, was there a method of cutting one and rearranging the polyhedral pieces into the shape of the other? (No melting allowed, only straight cuts.)

This problem was solved by Hilbert's student Max Dehn (1878–1952). Dehn proved that it was impossible. To understand why, we need to mention *invariance*. In mathematics, invariance is a property of an object that will remain unchanged after a transformation is applied to the object. The area of a polygon will remain unchanged if you cut it and move it around, so area is an invariant property of polygons. Hilbert's problem was asking whether volume was an invariant of polyhedra. Dehn discovered that there is an invariant quantity within three-dimensional shapes that is dependent on *dihedral angles* (the angles between two faces of a polyhedron). If this invariant is the same between two solid bodies, then you can cut and rearrange them to fit one into another. But if this *Dehn invariant* is different, you can't do with polyhedra the thing you can with polygons. The Dehn invariant of a cube is different to the Dehn invariant of a tetrahedron, so you can chop and move the pieces all you like, you can't reassemble them into the volume of the other.

It's beyond the scope of this book to go through all of Hilbert's problems, but we must mention another one. This was the eighth problem, still probably the most famous unsolved problem in mathematics today, and which most mathematicians regard as being as important as Fermat's last theorem. This is the *Riemann hypothesis*.

Formulated by Bernhard Riemann in 1859, the Riemann hypothesis deals with something called zeta functions and its zeros (its roots). A Riemann zeta function is the sum of a number series in the format $\frac{1}{n^s}$, where the n would start from 1 and go into infinity, and the power s would be some complex number. Recall Euler's Basel problem from Chapter 21, about finding the limit of the infinite sum of reciprocal squares: that's a zeta function where s is 2. But what if the value of s is a *complex number* – that is, a number made up of a real number and an imaginary number?

Things are, indeed, starting to get more complex. We saw that back in the Renaissance, imaginary numbers like $\sqrt{-1}$ were accepted as numbers although they don't make much sense in the real world, and nor do they appear on the number line of real numbers. How could they be worked with? We could refer to them as i for the purposes of equations but more was needed. As usual, mathematicians came up with a new system, adding a second perpendicular line to the number line for imaginary numbers only, and thus creating a number plane. On this number plane all the numbers, including those which have imaginary parts, could be plotted. This is called an *Argand diagram*, described and defined between two mathematicians, Caspar Wessel (1745–1818) and Jean-Robert Argand (1768–1822).

Such diagrams allowed mathematicians to represent numbers such as $z_4 = 2i$, but also a complex number such as $z = -2 - 4i$, with two parts, real (–2) and imaginary (–4i). The Riemann zeta function is very useful for investigating prime numbers, and Hilbert's eighth problem asked about them too. So the million-dollar question (literally: this is another of the Clay Institute's Millennium Prize Problems) is: For which values of s is the Riemann zeta function equal to zero? Remember that a Riemann zeta function is the sum of a number series in the format $\frac{1}{n^s}$, where the n would start from 1 and go into infinity, and the power s

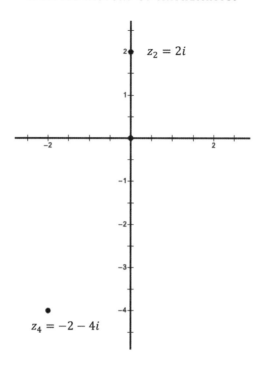

would be some complex number. No one has yet come up with the answer, so we shouldn't feel bad that we haven't either.

Hilbert was adamant that mathematics had the power to solve any problems that could be posed – as long as they were well-formulated. He studied Euclid's axioms and proposed and analysed twenty-one of his own. Axioms can be decided, and if the whole of mathematics could be organised along axiomatic lines, he said, it would be a decidable system. Every mathematical problem had a solution. His famous words, 'We must know, we will know!' are carved on his gravestone.

But the huge advances in mathematical and other sciences at this time had led some to start wondering about the limits of knowledge. As we'll see in the next chapter, others took a more philosophical approach to the subject, and one young logician was to shatter whatever hope Hilbert had for mathematics' ability to provide all the answers.

There are some things we just can't know with the tools we have at the moment. One of the oldest challenges in mathematical geometry that found a conclusion at this time was the classic problem of squaring the circle (Chapter 5). In 1882, Ferdinand von Lindemann (1852–1939), who was a teacher of Hilbert and a student of Klein, showed that π is a transcendental number. This means that π is not a root of any polynomial with coefficients which are rational (can be expressed as fractions). Numbers like π can't be constructed on a number line, and so, because a circle's area depends on π and a square's does not, the two are not commensurable.

Squaring the circle was one of the three problems posed by the Greeks that remained unsolved for many centuries, the others being to trisect any angle, and to double the cube. They were eventually settled – proved to be unsolvable – with the invention of abstract algebra in the nineteenth century, showing that not all numbers can be constructed. So for some problems, we just don't have the tools yet. In these cases we had to wait twenty or more centuries to find them. But the good news is that mathematicians kept and keep going, trying to solve old problems and coming up with new ones all the time.

One of the places they can do this is, thanks to Georg Cantor and Felix Klein, at the International Congress of Mathematicians. Here, mathematicians from around the world now come together every four years, and their numbers are growing, from just 250 mathematicians in 1900 to more than 3,000 a little over a century later.

There are so many participants now that you might say that David Hilbert, the star of the second ICM back in 1900, dreamed up a hotel just for them. This was a thought experiment Hilbert came up with to talk about the paradox of infinities. Say you're the manager of a hotel of infinitely many rooms, all numbered, and these are all occupied by guests – you are full. But here comes another mathematician attending the ICM and needing a bed for the night. Can you accommodate her? A normal hotel would have to refuse, but you have an infinity of rooms. Yes! But how do you tell her which room number she'll be in and direct her there? It's in infinity! Instead, Hilbert said, you can just tell everyone to move one room along, and then give your new guest the keys to Room 1. Mathematicians have had fun with this problem, devising ways to accommodate infinitely more people, even if they arrive concurrently on coaches, or even ferry-loads of coaches . . . There's mathematics for it all.

Perhaps we can imagine a book with infinitely many problems in it! There will always be space for more, infinitely more, in mathematics.

CHAPTER 30

In Pursuit of Perfection

'This was one of the great events of my life, as dazzling as first love.' So Bertrand Russell (1872–1970), one of the greatest philosophers of the twentieth century, described first encountering the mathematics of Euclid as an eleven-year-old boy. There is a cool and wonderful beauty in the *Elements*, the certainty of the structure of axioms and proofs that develop into a whole view of geometry. But as we've seen, in the nineteenth century mathematicians realised that alternatives also exist, in non-Euclidean geometries. As Russell continued reading and doing mathematics, he discovered that too. All of that certainty was in the past. Modern mathematics was the mathematics of different geometries, dimensions higher than three, infinities, sets and new abstract ideas. And that was even more exciting.

He went to Cambridge to study mathematics and stayed on as a lecturer afterwards. At some point he became interested in the *foundations of mathematics*. This is all about original premises, the definitions and axioms. As problems were identified in geometry – as we saw with Euclid's fifth postulate, then later in sets of numbers and the sizes of their infinities – mathematicians started to think about where the problems originated. And they found that they were inherent in the very foundations of mathematics! The question then arose: was there a way to work more on these basic notions – axioms, definitions, theorems – so as to clarify things a bit more and avoid future problems in mathematical sciences?

Russell thought about this. In May 1901, while working towards his first major publication, *The Principles of Mathematics* (1903), he came across something that bothered him. It's now commonly known as a *Russell paradox*, and it goes something like this. Say there is a barber who declares that he shaves everyone who doesn't shave themselves. That seems straightforward enough. But does the barber shave *himself*? If he does, he shouldn't, because he should only shave those people who don't shave themselves. And if he doesn't, he should, because he should shave all those who don't shave themselves. It is logically impossible. There's no way to solve this contradiction.

Mathematically speaking, this had big consequences. Russell was really talking about sets and how you define them, and further, the whole programme of reducing mathematics to logic that mathematicians such as Gottlob Frege (1848–1925) were engaged in. Russell's paradox rests on a distinction between types of sets. There are *normal* sets, which aren't members of themselves, and *abnormal* sets, which are. Take for example a set of squares in a plane. The set itself is not a square in a plane, so it's a normal set. But there is something called the *complementary* set, which is everything that surrounds the original set. In this example, the complementary set is abnormal, because that set is itself also not a square in a plane, and so is one of its own members.

So then, can we say that all sets are either normal or abnormal? No. What about the set of all normal sets? If that set were normal, it would be contained in the set of all normal sets – it would be a member of itself, and therefore abnormal. But on the other hand, if the set of all normal sets were abnormal, it would not be contained in itself, and so would therefore be normal. So the set of all normal sets is *neither* normal *nor* abnormal. It both contains itself and does not contain itself. This was Russell's paradox.

Around 1908, Russell offered a way to avoid this paradox by introducing his *theory of types*. It revolved around the problem of defining sets in the first place so as to avoid these sorts of vicious circles. So, rather than a set simply being an arbitrary collection of elements, and the existence of those elements themselves being sufficient for determining the set, Russell introduced new distinctions and hierarchies that would stratify elements of sets in order of their complexity. He gave new language to how we talk about elements and sets. For instance, you should say that some collection of objects are the same *type* if they all possess a certain property. He introduced hierarchies to prevent self-reference by introducing new rigour to the formulation of sets. Statements such as '*x* is a set' could not apply to itself.

This was a reworking of what became known as *naïve set theory* from the ground up. It was a huge undertaking and was feeding into something much bigger – which was to result ultimately in a hugely influential book that Russell started working on with his friend and former teacher, mathematician and philosopher Alfred North Whitehead (1861–1947).

While Russell was developing these ideas, another mathematician in Germany was coming at the problem of achieving a contradiction-free set theory from a different perspective. Ernst Zermelo (1871–1953) had studied in Göttingen and knew Hilbert and his twenty-three problems. In 1902, under guidance from Hilbert, he started work on the first problem, that of whether the continuum hypothesis can be proved or not (Chapter 29). To do that, you would have to find a way to enumerate the infinite sets of real numbers and natural numbers in order to establish their cardinalities. Zermelo got a bit stuck on all of this. He was missing something.

The genius of his answer to this conundrum is truly astounding. In 1904, Zermelo came up with a simple idea: introduce the *axiom of choice*. You can choose an element, or a number of elements, from your original sets and use them to construct new sets. And these would then have elements in common. Say you decide to pick every second element, or the elements which are to the far left and right of your sets. Whatever you decide, there is always a choice you can make without deciding how you're going to proceed further. That comes later, *how* to do things; first comes the knowledge that you can make a choice. Over the years, mathematicians worried that this wasn't quite good enough; there should be some kind of constructive idea about the type of choice you can make. But at the time, Zermelo used his axiom wisely. It was the basis for his proof that every set can be *well-ordered*. Some sets are well-ordered already, and some can be made well-ordered by making certain choices. Zermelo proved it!

There are some rules that must be followed for a set to be well-ordered. For every two elements, there must be a way of distinguishing which is smaller. If there is then a third even smaller element, that must be shown to be smaller than both the first two elements. In mathematical terms, if $a > b$, and $b > c$, then $a > c$. Finally, a well-ordered set is one in which every subset has a *least* element. That is, every set within the original larger set has to have a smallest element. In short, a well-ordered set is one in which its elements can be put in some kind of order. The set of natural numbers, for example, is a well-ordered set. The set of integers, ordered in the usual way, is not. That is because the set of negative integers doesn't have a least element. But we can have subsets with *least* elements if we make some wise choices!

Zermelo's proof of the well-ordering property made him famous, but his ideas were met with criticism from the mathematical community. He answered them in his 1908 paper *Neuer Beweis* (*New Evidence*) where he argued pointedly that even his critics used the axiom of choice when dealing with infinite sets! By this time Russell's paradox had been published, and this had encouraged Zermelo to publish his seven axioms governing set theory, one of which was the axiom of choice. He didn't himself manage to prove his axioms were consistent, but later mathematicians did, improving the system in the process. It eliminates the contradictions found in classical set theory.

Meanwhile, Russell and Whitehead were forging ahead in their own very abstract thinking about very simple things. Their monumental work on the foundations of mathematics appeared between

1910 and 1913 in the hefty three-volume *Principia Mathematica*. It shared its title with Newton's great book of 1687, but whereas that work dealt with mathematics to explain phenomena in nature, Russell and Whitehead's *Principia* aimed to explain all mathematics in terms of logic. Its authors wanted to analyse the premises of mathematics in order to try to reduce the number of definitions and axioms. They also wanted to use logic to solve the paradoxes that were plaguing set theory, as we saw earlier. It was, in other words, a book that was trying to present the most important principles of mathematics from purely logical foundations.

The manner of doing mathematics at this time is now often called *logicism*, and it was this that *Principia Mathematica* hoped to establish on a proper and unquestionable footing. Russell's logical work had a lasting and significant impact on major thinkers in mathematics and philosophy to follow. But it didn't quite achieve the goal of eliminating all the problems once and for all. Another mathematician, logician and philosopher now came on the scene to prove that mathematics was just another system with any system's inherent flaws.

Kurt Gödel (1906–78) was born in Brünn (now Brno, Czech Republic) and showed an early talent for mathematical logic. After completing his doctoral studies at the University of Vienna he applied himself to thinking about axiomatic systems. In 1931, he proved that the axiom of choice can't be either proven or shown to be false by the other theorems of set theory. (Zermelo felt he had already proved this important result himself, so this coup was not the start of what might have been an enduring friendship between the two.) But Gödel went further, and he had Russell and Whitehead's *Principia* in his sights. He had received a personal copy of the book and found it lacking in formal precision. He now put forward fundamental proofs to show that in *any* axiomatic mathematical system – and Russell's work had been precisely this, an attempt to put mathematics on an axiomatic basis – there are propositions that you can't prove or disprove (say they are true or untrue). He then also proved that a mathematical system – whatever it is – cannot prove its own *consistency* (in other words, the system always behaving in the same way). This is now called Gödel's *incompleteness theorem*. There are always true state-ments that can't be proved. This was a landmark finding in twentieth-century mathematics, indeed in intellectual thought altogether. It is the paradox at the heart of mathematics. It will never be complete, it will never be entirely consistent (for sure), it will never be finished. Rather than worrying about any kind of perfection, we should just keep on doing mathematics. The important thing is to keep perfec-tion in mind and try to get as close to it as we possibly can!

Russell, exhausted by the huge *Principia*, had long since stopped working on mathematical logic. He recognised that some of Gödel's criticisms had value, but had turned his attention to philos-ophy, which is what he is now best known for. He was a staunch pacifist, although the Second World War changed his view. He was awarded the Nobel Prize for Literature in 1950 for his many philo-sophical writings on humanitarian issues (he did write some fiction afterwards, opinions on the quality of which vary). In his very long life, Russell became a very public philosopher, and you can still find wonderful videos of him talking in his old age about how people should live their lives.

In Vienna, Gödel tried to bear up against the rise of the Nazi Party but, after a couple of nervous breakdowns and a fear that he would be conscripted into the German army, he left for Princeton at the outbreak of war. He had lectured there during the 1930s and now made it his home, turning his attention in his later years to philosophy and physics. At Princeton he soon became a very close friend of Albert Einstein, who was also based at the Institute for Advanced Study, and they took long walks together – one can only wonder what they talked about! Einstein had recommended Gödel for US citizenship and in 1947 accompanied him to the citizenship exam. The judge happened to ask Gödel if he thought something like the Nazi dictatorship could ever happen in a place like America. Gödel had already confided to Einstein that – of course! – he had discovered an inconsistency in the US Constitution that he thought could allow this to happen. No system was perfect after all. Thankfully he was made to stop talking just in time so that his citizenship papers could be signed.

CHAPTER 31

Ramsey and Friends

Think of the most fun parties you've been to. No doubt they'll have involved lots of friends, and perhaps some of them a few strangers as well, who you had interesting chats with. If you were trying to plan a party with a perfect mix of friends and strangers – say, where at least three people would know each other or wouldn't know each other – are you going to have to expect a houseful of guests? Not necessarily. Surprisingly, a mixed group of just six people is large enough for there to be three people who are all either mutual acquaintances or mutual strangers.

This example, or rather, the mathematics behind it, was developed by Frank Ramsey (1903–30), a brilliantly innovative British philosopher, mathematician and economist. It brought about a brand new mathematical theory, now called *Ramsey theory* – an entire branch of mathematics that has been accumulating, influencing and developing ever since.

It started in the 1920s, when Ramsey published his ideas on making decisions using logic in *On a Problem of Formal Logic*

(1930). There he also introduced combinatorics to solve a problem similar to the one we've just described (originally the problem was stated in slightly more obscure mathematical language, but later on its everyday example was formulated as we've just seen). Really this was an aside to Ramsey's main goal, which was to create a system that could be used when making decisions, including our own beliefs. Ramsey was struggling with questions involving logic and complex, chaotic mathematical systems. His hunch was that the sheer size of a system should be enough to bring at least part of it into some kind of order. He thought probability theory would also be useful there – in an earlier publication he had stated that in greater detail (*Truth and Probability*, 1926). But the friendship problem is what he is best known for today in mathematical circles.

Let's look at the example again and the steps by which it gets to its result. We're going to make some pretty pictures here – another reason why Ramsey theory is popular! Our diagram will connect our units (people, in this case). We'll represent the units as vertices, and the connections between them we will call edges. A *complete graph* connects all the vertices in all possible ways. So let's do that with our six vertices, making a hexagon.

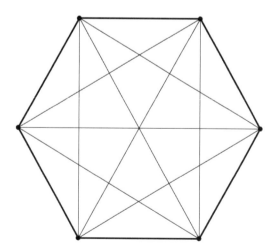

We can use different lines to represent the edges (or connec-tions): solid line for friends, dotted line for strangers. Here we've just added any lines to start off with (there are six friends and nine strangers). Now, if you really wanted to try out *all* the possibilities of the ways the lines could be combined, it would take you some time – there are 2^{15} ways of drawing this graph for lines of two types (or two colours)! Luckily we can focus our investigation. What we're interested in is to show that, even with a group of just six, *at least* three people will either be friends or strangers.

Let's look at what happens at one vertex. It has five edges coming out of it, a mix of solid and dotted lines (because this group of people is a mix of friends and strangers).

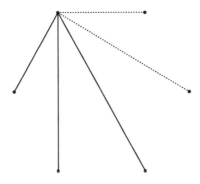

Now let's connect the other vertices using whatever edges we want:

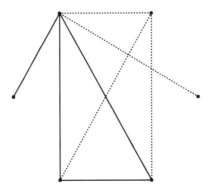

You'll see that a triangle has appeared, and we only need one triangle with edges all of the same type to solve the puzzle – here, three people know each other. We found a triangle quickly here, but this holds with *whatever* edges come out of the vertices: however you draw the two types of lines, however you choose to connect the vertices, you will always end up with at least one 'complete' triangle with the same type of edges. (There are in fact 156 ways that a group of six can include at least three friends or three strangers!) This is *Ramsey's theorem* – in its most basic form it says that in any sufficiently large complete graph, there will always be a relationship between vertices defined as we have above – connected by edges of the same type.

This neat little problem gives us our first example of a *Ramsey number*. A Ramsey number is the minimum number of vertices a complete graph can have which will always contain a group of vertices that are connected with one of two types of connections. And these complete subsets of the whole graphs we call *cliques*, a mathematical term that means something similar to its ordinary usage – a group of vertices that are connected in the same way. Ramsey numbers are written in the form $R(s, t) = n$, where n is the minimum number of elements in which s or t cliques will exist. So in our example, the Ramsey number is $R(3, 3) = 6$: six is the smallest number of vertices that a graph with two cliques (s and t), both of size 3, will have. Mathematicians have since expanded on this problem to look at different and larger Ramsey numbers.

Ramsey was an exceptionally talented man and mathematician, one of the greatest minds of the twentieth century. He still has a huge following not only in the mathematical world, but in philosophy and, especially, economics too. He published seminal papers in economic thought, so important today but in a sense tangential to what his work means for the development of modern mathematics. He died far too young, aged just twenty-six, and only a couple of years after publishing the paper that gave us the Ramsey theorem – a powerfully simple proof of an intuitively understandable problem with very interesting possibilities. Now, others took it up and developed it further.

One of the more famous mathematicians that worked on this was Paul Erdős (1913–96). He has to be one of the most prolific mathematical authors of all time, and certainly of the twentieth century. He collaborated with mathematicians around the world and wrote more than 1,500 mathematical papers with many different colleagues. Born in Budapest to two Jewish mathematics teachers, Erdős became very famous in 1933, when he was only twenty. He solved a theorem that had been posed in the nineteenth century, which stated that for every positive integer n, there is a prime between n and $2n$. Between 2 and 4 there is a prime: 3. Between 10 and 20 there is a prime (well, a few actually: 11, 13, 17), and so on. The theorem had already been proved by a Russian mathematician in the previous century for any numbers n and $2n$, but the proof was very complicated. Erdős, on the other hand, provided a simple and elegant proof. This was to become something of a trademark in his further work: he tackled seemingly simple but fiendishly difficult problems and came up with beautiful solutions.

Having fled Hungary following the Nazi takeover, Erdős went to the United States, flitting around university mathematics departments here and there. He couldn't settle. He was a brilliant scholar yet, even his friends admitted, a rather eccentric person. In the 1950s this came to have serious consequences. With the rise of McCarthyism in the United States, Erdős, like so many other innocent people, came under suspicion. They found that he had corresponded with a Chinese mathematician, and had made some honest but ill-judged comments about Marx's greatness and his plans to return to (now communist) Hungary to see the family who'd survived the Holocaust. Rather surprisingly for a mathematical scholar, Erdős also had an FBI record, having been picked up by the authorities with two other mathematicians near a military radio transmitter in 1941 – they had simply been too wrapped up in discussing mathematics to have noticed the no-entry sign! It was all used against him and he was not allowed back into the US until 1963, but by this time travelling around and staying with friends had become his lifestyle.

Erdős spent the majority of his life working on number theory and Ramsey theory. People say that the Ramsey *theorem* wasn't created by Erdős, but Ramsey *theory* was to a large extent due to him.

Ramsey theory is a branch of *combinatorics*, which is mainly about counting and finding the properties of finite structures in mathematics. A combinatorial problem should involve some or all of the following: enumeration (counting); recognising structures by some criteria; ways of describing a construction of some structure; and optimisation, or finding what is the best thing to do in a structure to get the best result.

Others worked on Ramsey theory too, as it can be applied to anything you can count. One of its more famous proven theorems is *Van der Waerden's theorem*, named after the Dutch mathematician and mathematical historian Bartel Leendert van der Waerden (1903–96). It's about arithmetical progression (where the same amount is added to a term to get each next term). It states that for any given positive integers r and k, there is a number N such that, if the integers are coloured or marked in r different ways, there will be at least k integers *in arithmetic sequence* whose elements are of the same colour.

Now, this sounds very unlikely – but so did the party puzzle with which we opened this chapter. Let's see what happens in this case. We'll use $r = 2$ and $k = 3$, meaning that we'll colour the integers in two ways (bold and not-bold), and look for three numbers forming an arithmetic sequence (where the same difference is added to a number to get the next one). Here we're trying to find the *minimum* number of ordered integers where this will happen: this is called W (standing for Van der Waerden's number). So let's see what happens when we colour our integers (bold or not-bold):

1 **2 3** 4 5 **6 7** 8

Nothing is happening yet. But if you add a 9, everything changes! If you make 9 bold too, you get three bold numbers that form an arithmetic sequence – 3, 6 and 9:

1 **2 3** 4 5 **6 7** 8 **9**

If you'd chosen not to make 9 bold, you can also spot a pattern of not-bold numbers that form an arithmetic sequence – 1, 5 and 9:

1 2 **3** 4 5 6 **7** 8 9

So either way, we have our sequence, and $W(2, 3) = 9$. This is a simple solution. Imagine now it being applied to much larger lists of numbers. The surprising thing is that there is a point somewhere in the counting that the most unlikely kind of order appears.

Another fundamental result of Ramsey theory involves the game called noughts and crosses or tic-tac-toe. No doubt you know how to play it; the aim is to get three in a row (horizontal, vertical or diagonal). If you're careful, you can always force a draw. This makes the game unremarkable for mathematicians. But say we play this game not in two dimensions, but three – on a cube. Say we keep adding boxes or cubes to our rows and columns, or play with more than two players, or play in four dimensions or higher. The Hales–Jewett theorem says that no matter how many cells or players or dimensions there are, you will still get a winning line (which we call a *combinatorial line*), provided you're playing on a big enough board. That's much more fun.

Ramsey theory tells us that you can't avoid things like this happening. It shows us that patterns are implicit in *any* large structure. Given enough elements – number series, stars, pebbles, whatever – one can always find an orderly substructure. In one sense, as a later mathematician said, complete disorder is impossible. Even if you complicate things so much that it seems there is no way that any kind of order could emerge, it does. And this is something that Ramsey theory taught us from the very beginning. But the complexity can be fiendish. Paul Erdős once said that, if a vastly powerful alien force came to Earth and demanded to know $R(5, 5)$ or they would destroy the planet, we should muster all the world's mathematicians and computer programs to solve it. But if they asked for $R(6, 6)$? Destroy the aliens.

Throughout his long mathematical life, Erdős was a constant source of new problems and questions that motivated not only his own work but that of his very large and wide circle of mathematician friends. As he got older he started appending humorous letters to the end of his name – we could say, he was coming up with definitions even for himself! He said he was a PGOM (Poor Great

Old Man) when he became fifty, LD (Living Dead) at sixty, AD (Archaeological Discovery) at seventy, and CD (Counts Dead) at eighty. He died aged eighty-three – a prime number, which would have pleased him.

Right until the end, he hunted out new and upcoming collaborators, and worked with an astonishing number of mathematicians around the world. As a result we now have an Erdős number. This is not connected to any mathematical proof, though he had come up with plenty of those. Rather it measures the 'collaborative distance' between a mathematician and the great man himself, and was introduced by his friends as a tribute to him after his death. So, Erdős himself has an Erdős number of 0. If you were lucky enough to have directly collaborated with him on a published paper, you have an Erdős number of 1. If you've published a paper with anyone who collaborated with him directly, but not with Erdős himself, you have an Erdős number of 2. And so on. Currently the biggest Erdős number is 13. In his lifetime, Erdős published papers with just over 500 different mathematicians, but based on the network of Erdős numbers, he's linked to more than 200,000 mathematicians around the world. This sounds like another problem for Ramsey theory!

CHAPTER 32

The Mother of Abstract Algebra

There is no rule that says you'll only become a famous or important mathematician if you are good at mathematics as a child. Emmy Noether (1882–1935) is certainly an example of that. She was the daughter of a mathematics professor in Erlangen, Germany, but aside from that she did not show any talent or interest in mathematics until she grew up. The only clue that she would perhaps be good at the subject was her skill at solving logical puzzles as a girl. She was always finding new ones to entertain her family and friends with.

Born to a Jewish family, Noether was the eldest of the four children, and the only daughter. Being female, she couldn't study at a university, so she was preparing to become a teacher of languages, having showed her proficiency in French and English. She passed her exams, but then the law in Germany changed, allowing women to attend (though not fully participate in) lectures at universities, as long as they had sought permission to do so from each lecturer. So Noether started attending the University of Erlangen, where her father taught, as one of only two women among nearly 1,000 students. First, she selected modules in modern languages, then history, and then mathematics. Suddenly, she discovered she was much better at mathematics than anything else and enjoyed it most of all her subjects! She started working on mathematics exclusively and finished her PhD in 1907 by the time full access was afforded to women. By then she was already known to mathematicians across Germany. She began to lecture at Erlangen, without pay, sometimes standing in for her father. Introduced to David Hilbert's work by a colleague, she published several papers extending Hilbert's methods, and Felix Klein and David Hilbert heard of her work.

Noether's doctoral thesis was on the invariance of polynomials: in other words, she was interested in the things that *don't change*, that remain invariable, when you do something to polynomials (expressions such as $ax^2 + bx + c$). She worked on this and on the symmetry in polynomials. Her ability to study such difficult and abstract concepts and her knowledge of invariance was of interest to Hilbert.

By this time, Hilbert was working on Einstein's theory of relativity. There was a problem there that Hilbert wanted to better understand and hence started corresponding with Einstein. In turn, Einstein, who had already published on the special theory of relativity in 1905, had realised its extent and began working on his general theory. The special theory of relativity stated that no material object can travel as fast as light. But the general theory of relativity (1916) eventually stated that gravitation is a very weak force and manifests itself only on massively large objects. Through it, very large-scale phenomena were able to be explained, including the movements of planets, black holes and more.

Although a physical theory, *relativity* has deep implications for mathematics too, in particular in relation to geometry: it changes our understanding of the curvature of space-time and four-dimensional space-time. We met with four-dimensional space-time in Chapter 28. Curvature is a mathematical way of measuring the way a line differs from a straight line. Hilbert wanted to understand in particular what was *invariant* in the theory of relativity, in a mathematical sense. If gravity as a force in the cosmos influences the shape of space, what is the mathematical description of this? We could change the frame of reference, but what would stay the same? Noether's work on invariance could help here, so Hilbert and Klein wanted her to come to Göttingen to work with them on this. Some members of the philosophical faculty initially blocked this move, refusing to countenance the idea of men being 'required to learn at the feet of a woman'. Hilbert fought back. He knew that Noether's contribution to mathematics was already more important than anything her sexist and anti-Semitic detractors had to say. In 1915 she started teaching at Göttingen, but didn't have a formal position; she had to lecture under Hilbert's name, not her own, and for some years.

Noether's coming to Göttingen proved to be incredibly important for physics and mathematics. She came up with a theorem that connected invariance with symmetry, now called Noether's theorem. There are quantities that nature preserves, for example momentum (a product of the mass and velocity/speed of an object) and energy. Energy doesn't come from nowhere; it is conserved constantly and transforms from one form to another. This is one example of the law of conservation. Noether connected the laws of conservation with the laws of symmetry. Every symmetry, Noether discovered, has a conservation law that is associated with it. Through symmetry, things are preserved in some very specific way. *Noether's theorem* tells us that every continuous symmetry of an action of a physical system (that is, a symmetry that can be viewed as a motion of some kind) has a corresponding conservation law. This now provided the field of applied mathematics with a different way of seeing, and very concrete mathematical descriptions connecting physical phenomena with mathematical symmetries. Noether's theorem has been hailed as revolutionary – as important for mathematics, some say, as Pythagoras' theorem was in antiquity.

Noether had an unusual gift for making connections between seemingly different and difficult concepts. Hilbert and Einstein were impressed by her discovery and wished for her to be given the recognition that she deserved. It was a long road for a woman. She finished her habilitation and was able to become an associate professor in 1919. But only in 1923 was she eventually paid for the work she was doing. Noether was one of the very first women to officially teach mathematics in Germany, at the University of Göttingen, and be paid for it.

She now went back to pure mathematics, and inaugurated a new area of study, now called *ring theory*. Rings, along with fields, groups and ideals, are elements of modern, abstract algebra. We've already explored groups (Chapter 24), so let's have a look at the others and Noether's work with them.

A *ring* of whole integers (positive and negative) is a good place to start. A ring is a kind of set where two particular binary operations occur. We can say that integers form a ring with addition and multiplication (but not division). This is because when we add, subtract and multiply whole numbers, we get results that are also whole numbers (they stay inside the group), but when we divide we yield at least some results that are not whole numbers.

Noether's most famous work is entitled *Idealtheorie in Ringbereichen* (*Ideal Theory in Rings*, 1921). This takes us a step further. Here Noether showed how an *ideal* can be considered. An ideal is a special subset of some ring. Ideals 'idealise' or generalise succinct characteristics of a subset of objects. One example is the even numbers. We say that they are an ideal of the ring of integers. The sub-group, or quotient group, formed by even numbers has a property (the most important property or characteristic) that they preserve: it is their *evenness* that remains invariant under the ring's operations. So, for example, adding, subtracting or multiplying even numbers will always yield an even number.

It appears from many accounts that Noether was of an incredibly generous spirit, and inspired many of her students and colleagues. Bartel Leendert van der Waerden, who we met in Chapter 31, was one. He learnt from Noether, became a kind of follower of hers, and promoted her work. He published a two-volume textbook in 1930–1 entitled *Moderne Algebra*, apparently written in the same way that Noether taught abstract algebra. It is the way abstract algebra is very much *still* taught in universities around the world!

Noether also became a very good friend of Russian mathematician Pavel Sergeevich Aleksandrov (1896–1982). Aleksandrov started coming to Göttingen at about the time Noether became a professor there. During one visit, Aleksandrov and Noether got to talking about the possible connections between topology and algebra. How does a topologist tell the difference between two things? One way is by employing something called *Betti numbers*, named after the Italian mathematician Enrico Betti (1823–92), who had come up with the concept. These refer to the number of holes on a topological surface. The surface of a sphere and the surface of a doughnut are topologically different because a sphere's surface does not have a hole, whereas a doughnut's does. This is described by these Betti numbers. A sphere, with no holes even if it is cut across, has a Betti number of 0; a torus, meanwhile, has a Betti number 2, as it can actually be cut in two different ways (sliced vertically or horizontally) and will produce two circular holes.

Discussing this with Aleksandrov, Noether came up with a new idea. She thought there was some algebra in here somewhere. As a consequence, we now connect a topological study of objects with some algebraic properties of such objects. Out of this conversation and friendship was born a branch of mathematics called algebraic topology.

Very soon after this breakthrough, however, the now dominant Nazi Party pushed through new legislation that removed all people of Jewish origin from their jobs. She and her brother, Fritz Noether (1884–1941), who was also a mathematician, would have to leave their places of work. Wisely they thought it best to also leave their country. Aleksandrov tried to find them both jobs in Russia, and Fritz eventually went there. Helped by US colleagues and an association that was formed to provide assistance to the dozens of newly unemployed and displaced German scholars, Emmy, meanwhile, took a position at Bryn Mawr College, a women's college in Pennsylvania. While there she also visited Princeton and saw

Einstein, on occasion giving lectures there too, though she felt unwelcome at what she called 'the men's university, where nothing female is admitted'. At Bryn Mawr she formed a group of doctoral students and postdocs dubbed the 'Noether girls', a number of whom went on to become successful mathematicians themselves. But it was not to last. Just two years after arriving in the US, Noether fell ill and died of complications after an operation on a tumour. She was only fifty-three.

The legacy of Emmy Noether lives on in mathematics departments around the world. She was something of a clarifier of the complex new theories then being introduced into mathematics, and she was able to inject algebra into all sorts of places, even the least expected. In some ways she was to modern mathematics what Picasso was to modern art. Rather than depicting what could be simply seen, he conceptualised the subjects of his paintings, depicting structures and relationships derived from experience. Noether was able to do the same for mathematics.

At her funeral, the German mathematician Hermann Weyl (1885–1955) gave a heartfelt eulogy. Weyl had also studied and worked at Göttingen and fled Germany earlier in the decade, becoming Noether's good friend in the US. He described her buoyant personality and rather wonderfully said that she was as 'warm as a loaf of bread'. Indeed, all the photographs of her show her with a wide smile, and she had many appreciative friends. Her legacy of beautiful insights into how different mathematical concepts could be connected together was also made possible through her own connections to many people from different places. Despite all the hardship and trauma she and her family lived through, her work remains a bright star that illuminates the study of mathematics to this day. She is fittingly called the mother of abstract algebra.

CHAPTER 33

A Fellowship Called Nicolas

On 30 August 1952, Nicolas Bourbaki was given an office at the École Normale in Paris. Although only eighteen, he was already a published author of mathematical textbooks and an influential figure in international mathematics. How had someone so young achieved so much, not to mention having landed a coveted office in one of the best mathematical schools in the world? Just where on earth had this Bourbaki come from?

He was a person who was, you could say, quite literally born of a pure love of mathematics. He also wasn't real. The imaginary Nicolas Bourbaki had been conjured up by a group of young mathematicians in the 1930s. How that happened is quite the tale.

In 1934, some French mathematicians working in Strasbourg, just over the French border with Germany, had looked with envy at what their colleagues in Göttingen like Hilbert, Noether and van der Waerden had. Could they form a group like that? They were all dissatisfied with the state of mathematics in France. After the First World War, which had dramatically reduced the ranks of French mathematicians, the subject was languishing. Not enough teachers, and no new textbooks. They were having to teach from the old books, and these books were severely lacking.

So, in 1934, three friends – André Weil (1906–98), Auguste Delsarte (1903–68) and Henri Paul Cartan (1904–2008) – started talking seriously about writing a new textbook together. A book that would be different from what was then on offer in France. It would be succinct, clear and would teach mathematics in an easy and constructive way. They enrolled others in their group too, including Jean Dieudonné (1906–92) and Szolem Mandelbrojt (1899–1983), and started meeting regularly in a little café in Paris's Latin Quarter.

But what to call themselves? Legends swirl around the origins of the group's name. One of them tells of a strange lecture at the École Normale Superieur in Paris, of which nearly all the members of the group were alumni. There was a tradition of student pranks there, and one of these seems to have struck a chord. It was a lecture to the first years delivered by a mysterious, bearded mathematician, very glamorously dressed. He went up to the board and wrote, 'Theorem of Bourbaki', instructing the students to prove what he was about to write. The students scratched their heads. The ideas presented were complete gibberish and in fact meant absolutely nothing. But who would know? Mathematics was becoming so abstract, very few people were able to distinguish between real and fictious theorems, real and fictious mathematics, even real and fictious mathematicians. For the speaker wasn't a mathematician at all, but a third-year student, and Bourbaki had been an unremarkable French military general of the previous century.

It was a prank, that's for sure, albeit a very abstract and intellectual one. Weil loved these sorts of jokes, and it sparked something for the little collective. They had already discussed that it would be easier if they were known by a single name. That person would sign themselves as the author of their book, even if behind the scenes the whole group would be contributing. But if it was to be the actual name of one of the group, wouldn't that potentially lead to rivalry and jealousy later down the line? In the summer of 1935, the group met together in the French countryside to hold their first conference. Taking a break, they went skinny-dipping in Lac Pavin when someone started exuberantly shouting 'Bourbaki! Bourbaki!' Suddenly it was decided: the person to represent them would be called Nicolas Bourbaki. And these playful, clever, somewhat mischievous mathematicians didn't just stop at the name. They invented an entire character with a rich backstory. Nicolas Bourbaki was to be a mathematician of Greek origin, a former professor of a (fictional) university, who now makes his living playing cards in Paris, having lost both his job and most of his fortune. Weil actually gave all this information to the editor of the journal that he submitted the group's first paper to in the autumn of 1935, saying that he had met Bourbaki in a Paris café, and that Bourbaki had passed the ideas onto him!

Soon, French mathematicians who would become some of the most famous and influential of the twentieth century joined the ranks of the Bourbaki collective. Starting from the first volume published in 1939, they wrote, over many years, an influential series of books called *Éléments de mathématique* (*The Elements of Mathematics*, though the French title has the unusual singular *mathématique*, signalling the group's belief in the unity of mathematics). Bourbaki's original plan was to write a textbook on *analysis*, a branch of mathematics dealing with functions, limits and calculus; it's about the methods and procedures used for analysing mathematical objects. But they soon realised that they needed to include other areas of mathematics, and duly expanded their concept, starting with set theory as a foundation. And, just as in the *Elements* of Euclid, the mathematics presented in Bourbaki's *Elements* is based on the axiomatic model, building from simple mathematical statements to quite complex abstract mathematics. Proofs, they said, had to be given, and done elegantly and correctly. There had to be clarity and rigour and abstraction. They wanted to put mathematics on a fresh, solid foundation that reflected modern, up-to-date knowledge. Their reasonably short, hierarchically organised mathematical textbooks would mean that no one wanting to study mathematics would need anything else.

Bourbaki's original impetus to improve the teaching of mathematics was carried forward by Jean Dieudonné, one of the group's founding members and its scribe; he would take care of the writing and finalising of the texts ready for publication. He published a number of books aimed at teachers at France's lycées to give them the necessary background to teach those pupils who would go on to study mathematics at university. Some say these are rather on the sophisticated side! He also helped to inaugurate what was to become known as the European modern mathematics movement in the 1950s. This was supposed to replace all the old mathematics that was being taught in schools, and introduce a new curriculum, for all children across the world, based on set theory and Bourbaki's conceptualisations of the 'mother structures' of mathematics. Dieudonné was exceptionally well connected, a member of various international bodies and a huge network of mathematical colleagues and friends, who all helped the modern mathematics movement get off the ground. It was rolled out in schools across the world. But the dream of a new model for mathematics education eventually foundered. Set theory might be easy for professional mathematicians, but it's not actually easier to teach than the old-fashioned mathematics. And as teachers were themselves not always trained in set theory, mistakes were made, which were difficult to rectify later on. If you learn something in mathematics that is factually wrong, it is difficult to 'unlearn' it, especially if you have misunderstood the basic premises.

Bourbaki insist on a very formalised language of mathematics – but that doesn't mean that as a group of mathematicians they are also formal! Far from it. Though their mathematics is very serious, Bourbaki's members are not, and humour has from the very first been a kind of glue that keeps the group together. From the witty beginnings of their name they have continued to use humour in and around their work, and funny stories about the group abound. What we know of their early conferences, they were famously chaotic – Dieudonné himself said they were like gatherings of madmen, with everyone shouting over one another. They named the conferences things like 'The Extraordinary Congress of Old

Fogies', referring to anyone over the age of thirty. They introduced a new 'dangerous bend' symbol, which appears in their papers whenever a particularly difficult idea is about to be introduced. Bourbaki applied for individual membership to the American Mathematical Society, thinking they weren't known in the US; they were, and they promptly got a letter back asking them to pay the higher price for institutional membership. They announced that Bourbaki's daughter, Betti, was to marry a lion hunter, and sent out wedding invitations. As none of the mathematicians from the original group were still working within it in the 1990s, someone announced that Bourbaki had died (the members have to leave when they are fifty).

But Bourbaki lives on. It is still a prestigious and very secretive group, with its actual membership unknown. (It is only after members have retired that they are able to talk about their involvement.) Though Bourbaki doesn't operate with the same regularity as it used to, members do still organise conferences every so often. There they agree on a programme of work, debate the material they have written, and around six or more drafts are made before things are published. The structure and focus of their *Elements* have evolved over the ninety or so years since it first appeared, encompassing the new mathematics that have been discovered and developed.

One of the most famous of Bourbaki's members after the Second World War was Alexander Grothendieck (1928–2014). He developed something called *category theory*. It built on the work of Noether and Alexandrov on the theory of structures, which had been introduced into algebraic topology in the middle of the twentieth century (Chapter 32).

Category theory can be used in (almost) all areas of mathematics. There is, then, a *category* we need to define. A category is formed of two types of thing: *objects* and *morphisms*, the latter of which are structure-preserving maps. These don't have to be real maps! They just show some kind of relationship between a source and a target. A typical diagram from category theory would be one with three objects, X, Y and Z, and their morphisms, which are relationships that connect them, f, g and $g°f$. (That little empty circle between g and f means there's a relationship between them – a combination of the two relationships, whatever they may be.) This can be made to apply to many things, so the objects and morphisms would have different meanings.

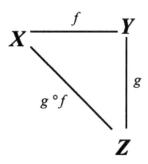

Sometimes people complain that mathematics is very abstract, and it really can be. This might be a good example of that! You may feel the itch here or at other points in the book to want to relate the mathematics to real life in order to properly grasp it or its importance. So, we can say that category theory is used in computer science and theoretical physics. But the whole purpose of abstract mathematics is to work out the laws that exist in general and with *any* entities, not to apply it to specifics. Why would you bother doing anything that relates to nothing and everything? Well, every abstract theory of mathematics finds an application at some point. But its real value is in its abstractness.

Bourbaki's greatest contribution is something like this too. Their strength and importance of their work is in the fact that it is a formal, clear presentation of the latest mathematical topics, and an up-to-date one at that. It keeps the axiomatic legacy of Euclid alive. (Interestingly, there was a theory that emerged in France in 1959 claiming that Euclid perhaps wasn't a real person either, but may have been a group of mathematicians writing under his name!) Although members of the group have been internationally prominent and Fields Medal winners, Bourbaki reminds us that in mathematics it is the activity itself that is really important, rather than a claim to individual fame or fortune; and that the best mathematics nowadays is very often a collective endeavour. So if you ever find yourself in Paris with an opportunity to go into the École Normale, pop along to see the office that still bears the nameplate *Nicolas Bourbaki* on the door. You can even write to him if you so wish – though a reply is not guaranteed.

CHAPTER 34

The Games People Play

Two members of a criminal gang commit a crime. They are caught by the police and held in separate cells. The police know they've done it, but don't have enough evidence to prosecute fully, so if they can't extract any more information from the prisoners then they will sentence them on a lesser charge to one year in prison. But then the police offer each individual prisoner a deal: testify against the other person, and you will go free, while the other person will get three years in prison. There's a catch though: if both testify against each other, they will both receive a two-year jail sentence. We could draw a matrix such as this to see all the possibilities.

Now, the prisoners don't have long to think about this, and there is no way to learn what the other will do. But they do know their partner has been made the same offer. There's no loyalty in this criminal gang – each prisoner is looking out only for themselves. What would you do in a situation like this? And what do you think is the best strategy?

This is a famous game known as *the prisoner's dilemma*. It was developed in the 1950s by mathematical scientists working for RAND Corporation, a research institute in the US. It was the height of the Cold War, and RAND were engaged in political-military wargaming, strategising approaches to decision-making in the shadow of international nuclear threat. RAND's thought-experiments were heavily informed by *game theory*, which concerns the mathematics of strategic interactions. Game theory was pioneered by mathematician John von Neumann (1903–57) and his Princeton colleague, economist Oskar Morgenstern (1902–77). Here's how that happened.

John (originally János) von Neumann was born in Budapest to a Jewish family. His father, Max, was a successful banker and his son acquired the 'von' title because of Max's contributions to the Hungarian economy. As a boy von Neumann had a prodigious memory and early on showed an exceptional aptitude for mathematics, publishing his first paper before he even went to university, where at the insistence of his father he studied chemistry – this being felt to lead to a more lucrative profession! He nevertheless continued with his mathematics and after receiving his doctorate lectured in Hamburg and Berlin, and was a post-doctoral student under David Hilbert at Göttingen. By his mid-twenties he was already famous in the international mathematics community as a mathematical genius. With the political situation deteriorating in Europe, he accepted a mathematics professorship at the new Institute for Advanced Study in Princeton in 1933, where he would remain for the rest of his life. During the Second World War he worked for the Manhattan Project, building the atomic bomb for the US military.

Von Neumann was also very much a man of his time mathematically speaking. He had an original interest in set theory and the foundations of mathematics. Independently from Gödel, von Neumann came up with the proof of what is now known as the second incompleteness theorem (Chapter 30). It shows that if you try to define a set of rules that include the statement that every rule can be proven to be either true or false, that rule must be false. Which means that every system has an inconsistency – in terms of logic, it is incomplete.

Oskar Morgenstern was a German economist who met von Neumann at Princeton. He came from an illustrious family, being the grandson of a former German Emperor and King of Prussia. He studied political science and became a professor of economics in Vienna. He was visiting Princeton in 1938 when Austria was annexed by Nazi Germany, and he decided not to return, remaining in the US for the rest of his life. Morgenstern was interested in cooperative games and economic *equilibrium* (a state in which opposing influences or variables are balanced), and wrote a paper on this in 1935. Upon reading it, his colleague Eduard Čech (1893–1960), a Czech mathematician, pointed out that another paper from 1928 had described something similar. This was on the mathematical analysis of parlour games, or games of strategy (*Gessellschaftsspiel* in German), written by none other than von Neumann in 1928. Parlour games can take many forms, and are usually about logic.

Morgenstern and von Neumann now decided to collaborate and develop their game theories further. They were greatly interested in bringing about an end to the Second World War, and in never having a war again. They started thinking about strategies where equilibrium is achieved and then maintained. Can you get to a theory that provides a mathematically precise formula to achieve this? That, in a nutshell, was what they wanted to achieve, and they published their results in their groundbreaking book, *Theory of Games and Economic Behavior* (1944).

Let's return to those prisoners and the dilemma they faced – cooperate by staying silent, or defect by testifying? At first glance you might think that cooperating is the best strategy with the least risk: imagining yourself into their shoes, you both serve some time behind bars but at least it's less than if you both testify. But let's look again, with our game theorist mathematician's hat on this time. Each prisoner is a rational player who is going to look at the mathematically defined outcomes in order to maximise their own return. So, you're going to work through your options. The choice is strictly to cooperate or defect. If the other person cooperates, you should definitely defect – you will be free! If the other person defects, you should also definitely defect; if you cooperate, you'll serve three years in prison, whereas if you defect, you'll only serve two. Mathematically then, game theory tells us that the best outcome for a rational player is never to cooperate, and to always be a defector.

But is this really what happens in the real world? Imagine if it did – if we were all completely self-serving there wouldn't be very much left of society to speak of. And that's not what we see around us. The difference between the prisoner's dilemma and what happens in the real world is that, in reality, games don't finish after you play one round. The playing, so to speak, continues, and you have to play the same or a similar game over and over again. If, evolutionarily and socially speaking, your aim is to survive, or in other words, to never lose catastrophically badly, then your dominant strategy is going to be very different. Rather than winning once but being forever after labelled as a defector, our lifelong strategies are based on things like kinship, group belonging, reciprocity and altruism.

Game theory examined this too! It showed that you can put numbers on things no one had thought of doing before in order to mathematically evaluate the best outcome. J.B.S. Haldane (1892–1964), an English-Indian mathematician-biologist, joked sometime in the 1930s that he would jump into the river to save two brothers, but not one, or to save eight cousins, but not seven. This is surprisingly specific! Haldane pioneered the application of game theory to evolution, and here he was talking about the mathematics underlying altruistic actions that are evolutionarily rational – family members are more or less genetically similar depending on the proximity of the relationship, so offering your life to save another is really rescuing copies of your own genes. Game theory can show how and when cooperation or altruism are dominant strategies and when they could and do take place. It makes the thinking behind decisions and strategies mathematical and measurable.

Though an early version had appeared in his 1928 paper on parlour games, only after working with Morgenstern on *Theory of Games and Economic Behavior* did von Neumann revise and expand what we now call the *Minimax theorem*. This involves a finite, two-person, *zero-sum* game – the result is going to mean a gain for one player and an equivalent loss for the other (we then say that the net gain of the game is zero).

We can describe the payoffs for the two players using mathematical functions, and these functions, when plotted on a graph, produce curves. Two plane curves combined will define a surface. Our task then would be to look for the most stable point on that surface. There is a point on this surface that is a maximum of one curve, and the minimum of the other. It turns out that there is always a value for two-person games such as this, where a stable point occurs – and this is the Minimax theorem. It's so called because it identifies a point at which the maximum value of the minimum expected gain for one player is equal to the minimum value of the maximum expected loss for the other. And as one person's maximum is another's minimum, there is always such a point on the 'saddle' of values where the two intersect. The point of intersection is the minimax value – the minimum for one and maximum for the other set of values. If you're losing, it's at this point that you minimise the worst-case scenario of yourself losing. The Minimax theorem holds that the strategy of both players realises this equality.

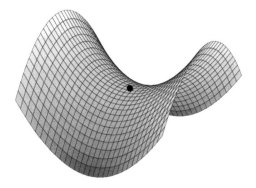

From the 1950s onwards, game theory became a huge hit with mathematicians, and it has been an enduring area of study ever since. Explorations into how the games could be expanded and what this would mean for the strategies and mathematics behind the theory abounded.

For instance, the prisoner's dilemma was extended in what sounds like an exceedingly fun mathematical experiment in 1980. Mathematicians were asked in advance to submit what they thought would be the best strategy to use for a two-player prisoner's dilemma game that was played indefinitely (or for a very large number of times). Many mathematicians submitted their strategies and were then invited to test them out in person at a tournament, where they would play using only that strategy. The winning one was also the simplest: tit-for-tat. It was submitted by Anatol Rapaport (1911–2007) a Russian-American mathematician and psychologist. The strategy is this: if you cooperate in this game, I will cooperate in the next. If you defect in this game, I will defect in the next. But as soon as you show goodwill and cooperate, I will do so as well. This tournament was run again, and mathematicians came up with many more new strategies. Which one won this time? Again, it was Rapaport's tit-for-tat! The repeated interaction mitigates against any feelings of hostility. There was one major flaw, though: the huge consequences of making a simple mistake. If you defect when I have previously cooperated, even if you didn't intend to, that hurts. Errors destroy trust, and mistrust destroys cooperation. The payoffs therefore are greatly reduced. It means that if you play fair, you will have the best chance of winning overall – but you must be extraordinarily careful to not make any mistakes!

Another very great mathematician expanded game theory in a different way, by looking at games of more than two players. This was the American mathematician John Nash (1928–2015), whose life was the subject of the film *A Beautiful Mind* (2001). He was very gifted mathematically, but struggled socially and, later, with his mental health. He too found himself at Princeton studying for his PhD just after the Second World War. He didn't attend lectures or even apply himself to reading textbooks, instead tackling problems by himself and in a completely original way. A paper he published at this time, on his development of game theory, would nearly fifty years later win him the Nobel Prize in Economics.

Nash looked at the strategic interactions of multiple non-cooperative players in a finite game. In this advanced game theory, things rapidly get more complicated. A player's outcome depends not only on their own strategy and decisions but those of (many) others. Nash's insight was that you can't predict what others are going to do in isolation – you need to understand and analyse what those players expect *others* to do and how they will act accordingly. This game theory is therefore about *all* the games one can play and are being played, and the mathematics (advanced probability theory; discrete mathematics, looking at graphs and networks) that underlie it. Nash made sense of the chaos of interactions and showed that the Minimax theorem is valid for any game (not just von Neumann and Morgenstern's zero-sum games) with a finite set of actions. This breakthrough finding, published for the first time in a 1950 paper only two pages long, was what we now call the *Nash equilibrium*. In all such games there is at least one Nash equilibrium, where each player's mixed strategy maximises their payoff against those of the others, provided the strategies of others are not changing.

It is perhaps not surprising that game theory really took off in an intense period of world war and the subsequent US–Soviet nuclear arms race. Questions of strategy and decision-making were of crucial importance for international peace. Such mathematically precise ways of predicting interactions, behaviour and outcomes, and the theory behind the ways equilibrium can be reached where no player loses catastrophically, were indispensable. Game theory has since become embedded in a wide variety of disciplines and real-world situations, from currency crises and climate change to traffic flow and penalty kicks. John Nash once opened one of his lectures on game theory by asking the students, 'Why are you here?' Well we might all just answer, 'Because of game theory.'

CHAPTER 35

Cauliflowers and Coastlines

In Chapter 33 we met briefly with Polish mathematician Szolem Mandelbrojt, one of Bourbaki's founding members. He was born in Warsaw but studied in France and settled there, becoming a professor of mathematics at the prestigious Collège de France in Paris. The Mandelbrojt family were Jewish and, back in Poland, were facing violent anti-Semitic persecution and the looming Nazi threat. Szolem helped bring some family members to France in 1936, one of them being his nephew, Benoit B. Mandelbrot (1924–2010).

After the Nazi occupation of France in 1940, life was perilous. Benoit couldn't attend school, so largely taught himself. It gave him, he said, an advantage in being able to see mathematics in a fresh and unique way. It helped, too, that Szolem had vowed to look after Benoit's mathematical education, and the two were always exchanging little mathematical problems to work out, challenging the other to find something more difficult next time.

In 1945, Szolem told Benoit about something his friends from France had worked on twenty-four years earlier that he couldn't really understand well. In fact, he said, no one could work out quite what was going on. If you can find anything new there, his uncle said, you will have your career made! The thing that Szolem was alluding to was about iterative functions and had been posed by Gaston Julia (1893–1978) and Pierre Fatou (1878–1929). Iterations (repeating a rule over and over again) of analytic functions generate a set of values (now called Julia sets) through a repeated procedure. Something was happening but it wasn't very clear what. Benoit looked at this, but couldn't find anything new or interesting about it. Calculating the values took an incredibly long time and no pattern was to appear or become clear through work like that. The time was not right for a new discovery, though Benoit couldn't then know it.

After finishing his PhD on 'Games of Communication' in 1952, Mandelbrot didn't tread the usual path towards a university professorship. He bounced between France and the US (at one point being a post-doctoral student of John von Neumann), taking up various positions in academia and industry, especially in the burgeoning field of computer technology, where IBM gave him a long and fruitful home. Over an astoundingly and intentionally wide-ranging career, he worked as an engineer, mathematician, computer scientist and financial markets analyst. Mandelbrot was a truly modern philomath.

But pure mathematics never left him. He still kept up the old habit from his childhood days with his Uncle Szolem, playing with mathematical problems in his spare time. Two things, seemingly unconnected, started to interest him. One was the iterative problem of Julia and Fatou – nether Benoit nor anyone else could work out what was going on. The other was a geometric problem involving plane and space-filling curves. The Italian mathematician Giuseppe Peano (1858–1932) had, all the way back in 1890, come up with a curve, now called the *Peano curve*, that, if iterated to infinity, would fill a plane. How can a curve (or *line*, in mathematical terminology) fill a plane? Well, it twists and turns in such a way that, if you repeat that again and again, it would come to fill an entire two-dimensional plane. This line will, being as squiggly as it is, reach all the places in the plane and fill it up that way.

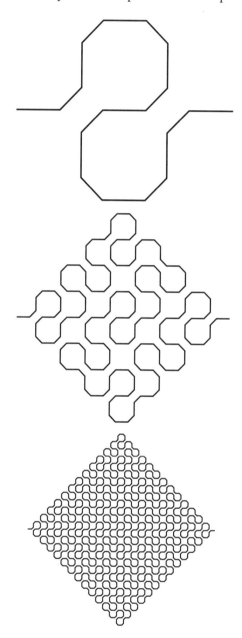

Now that he was at IBM, with access to the latest developments in computing technology, Mandelbrot returned to that iterative problem in 1980, thirty-five years after his uncle had first mentioned it. The Peano curve gave him an idea to try something new – do many iterations and see what gets plotted on a graph. He designed some of the first computer programs that would be able to print graphics, plugged the mathematics into the machines and bam! There was a noticeable *cloud* of solutions that seemed to come up again and again when you iterated very simple functions. He then tried with a very simple equation, iterated it, and this is what came up:

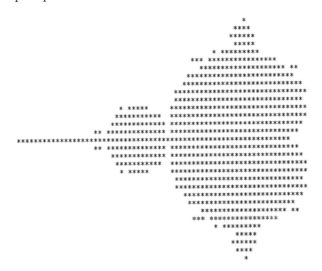

All the computer-aided results seemed to form a cloud-like shape. What was that simple equation? It was $f(z) = z^2 + c$, where c is a complex number and z are the iterative values. You start from $z = 0$, and then keep putting your results back into the equation, repeating it many times. The results seem to cluster, and if they do that in a particular way, we say they are in the *Mandelbrot set*.

The Mandelbrot set is very much about complex numbers. As we saw in Chapter 29, complex numbers 'live' in the complex plane: the horizontal axis gives us real numbers and the vertical the imaginary, and combining the two creates a plane of complex numbers. But what we didn't look at earlier was the *distance* between the point denoting a complex number and the origin of the complex plane. This is a segment that goes from the coordinate origin to the complex number's point.

It turns out that whatever complex number you choose to start your iterations from in the function $f(z) = z^2 + c$, there are only two possible outcomes. One is that the resulting points rapidly get infinitely large in size (and distant from the origin). The other is that all the resulting points remain in the complex plane within a distance of 2.

If you tried to do all this by hand it would take a hugely long time. This is why the use of computers was absolutely essential to the mathematical understanding of *fractals*, the new term coined by Mandelbrot from the Latin *fractus*, meaning fragmented or broken. You may have heard of – perhaps been scared by! – this mathematical term before. A fractal is a particular mathematical structure that is infinitely complex. It has different dimensions to those we are used to (and we'll get to that very soon). It is essentially an eternally repeating pattern which, if you zoomed all the way in or all the way out, would always look pretty much the same. There are many very lovely images of fractals available online, with the iterations in different colours, and videos too, which show the self-repetition when you zoom in. We've all seen fractals (strictly speaking, partial fractals) though. Think of the Romanesco cauliflower, made up entirely of self-similar repeating buds; if you cut its florets from the main vegetable, they too would resemble the whole cauliflower.

This was one of Mandelbrot's great insights. Having provided the general theory of fractal properties, he noticed – because he had noticed fractals – that fractals were almost universally found in nature. And that was worthy of mathematical study. This was an abrupt change from the increasing abstraction of the subject which had dominated mathematics between the 1870s and 1930s, when Mandelbrot had first cut his teeth in the mathematical community of his uncle. Mandelbrot described this once as the break between mathematics and reality. He thought that this was the issue that plagued mathematics – and mathematicians. No one outside of a small, hallowed circle could fully understand what was going on. He wanted, instead, to turn things around, and look at the mathematics of objects that resemble nature.

These objects were messy and rough. And this was another bone that Mandelbrot wanted to pick with mathematical physics. For too long the effort had been to find 'smooth', precise paradigms to describe nature. But they couldn't do it, not entirely – the theories were incomplete and in some cases radically flawed. In mathematics, precision is usually exactly what people are interested in; the whole idea is to tie up all the loose ends. But for Mandelbrot, the messiness was key. He had realised there is a kind of *roughness* to fractals, and that fractals differ from each other in some way. But how messy are fractals? And how can you best describe things that are messy? Can you put a number against their messiness? Mandelbrot was working on the whole mess phenomenon, so to speak, and discovered a special kind of order in it.

Mathematics wouldn't be mathematics if it wasn't about finding some general rule and some kind of measurement even in the messiest of places. In the 1960s, Mandelbrot turned his attention to some empirical data that had already been collected on the geometry of the British coastline. How could you measure its length precisely? Imagine looking at the coastline on Google Maps, zooming in more and more. What appeared once smooth and straight is anything but. Then imagine you have an ability to zoom in even further, so you can see every nook and cranny in the wiggly line of the coastline. It is an incredibly complicated curve. In reality, the coastline border is finite, but if you try to measure it, its length will tend towards an infinitely large number.

Mandelbrot came up with a new concept: the measure of roughness of a geometrical object. There was an order in this chaos that became apparent to him. And that order *was* the measure of messiness or roughness that he wanted to explain mathematically. To do so, he used the findings of another Jewish mathematician, the German Felix Hausdorff (1868–1942). Hausdorff was interested in problems of measure and in 1919 he came up with a number that allowed dimensions to be a given a value of a fraction. Dimensions, in Hausdorff's idea, can be represented through the relationship of some magnitude and the number of whatever dimension in which it finds itself. We all know what two- or three-dimensional things look like, but what about a thing which is in a $\frac{2}{3}$ dimension? This may seem confusing at first, but let's see how it works.

We start, on the left, with magnitude, which is first a line. In the next dimension up (the second dimension, or plane), that line becomes a square, and then in the next (third) dimension, a cube.

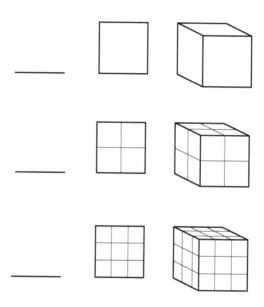

Note that in the top example the line is not divided, and produces an undivided square and an undivided cube. In the second, the line is divided into two, and produces four squares (dimension two) and eight cubes (dimension three). And so on. This ability to describe the original divisions in low dimensions that lead to much more complicated shapes in higher dimensions meant that a number could be placed on a fractal's measure of roughness.

And here was born the art of roughness. When Mandelbrot first mentioned that he used Hausdorff numbers to describe his measure of roughness to his mathematical friends, they told him it was a silly joke. Hausdorff numbers didn't really mean anything, they are not real things, he was just playing around. But using this formula has enabled the study of roughness through the division of the original length one uses to create higher-dimensional objects. It means we can create smoother and rougher fractals depending on how many divisions we choose for the original line, which translates into the divisions in higher dimensions.

The fractal structures' rules of geometry are, as we have seen, often quite simple. You get a formula, and you keep repeating it, and you end up with some mathematical image or model that closely resembles something from reality. But fractals are also very complex and very useful. Using fractals, computer-generated graphics can be made to very much resemble the roughness of real objects. You can create a surface or a curve which is a fractal by specifying its roughness, leading to, among other things, the astoundingly realistic effects we see in CGI films and video games today. You can also use fractals in medicine, as there are many examples of fractal structures in the human body – a human lung, for example, has a small weight and volume, but a huge and very complex surface area. Fractal geometry can model this.

Someone mentioned, after Mandelbrot died, that his ambitions were of Keplerian proportions. He wanted to find new, different ways of describing the real world through mathematics. He certainly achieved that in his work with fractals, and their huge applications to art, computer science and the mathematical modelling of structures that appear in nature.

The Unfinished Business

In the early 1960s, a ten-year-old boy picked a book off the shelf of his local library. It was called *The Last Problem*, written by historian of mathematics Eric Temple Bell, and was about the resolutely unproved, possibly unprovable, Fermat's last theorem (Chapter 18). Here the boy learnt about a scribble in the margin of a book, written by someone who was primarily a lawyer, more than three centuries ago. Who would've thought that this would have been a thorn in the side of professional mathematicians who had, ever since, tried to prove or disprove it? Fermat's last theorem – which stated that there are no three numbers (that are not zeroes) that can all be put to a power greater than 2 such that the sum of the first two numbers would equal the third – tormented the world of mathematics, and justifiably so. A child could understand what it meant, and yet even after all the complex mathematics that had appeared since it was first stated, no one could prove it one way or the other.

That child who was now initiated into the theorem's mysteries was Andrew Wiles (born 1953), who would in time become one of the best-known English mathematicians. Wiles was hooked. He understood the problem, he could see that it *looked* simple, yet none of the famous mathematicians over the centuries had been able to make any conclusive headway. He decided then and there that he would try to solve it himself. Mathematicians love a challenge! And sometimes people *become* mathematicians because of challenges like this.

And so began a life's work which, in the course of it, would also bring about quite a lot of new mathematics. Wiles studied undergraduate mathematics at Oxford and did his PhD at Cambridge – not on Fermat, as it turned out: it was deemed too risky to spend years on such a problem and get nowhere. Wiles became a professor of mathematics at Princeton in 1982, and the last theorem continued to bubble away for him in the background. For some time he'd realised he needed something else to move his thinking on Fermat forward, but he wasn't quite sure what it was.

Meanwhile, the huge advances in and availability of computing technology had revolutionised areas related to mathematics. Could computers cast their magic wand over the last theorem too? A program could now simply run and check, from the choice of millions of numbers, whether the sum of two numbers would equal the third when they are all raised to a power greater than 2. That way, you might think, we could straightforwardly check whether what Fermat had stated was true or untrue. But there would still be infinitely many numbers left! Compared to infinity, even 300 million numbers is but a jot, and certainly not sufficient to persuade mathematicians that the matter was settled. That was why a mathematical proof was still necessary.

Even with all the twentieth century's remarkable technological breakthroughs, it all boiled down to the same deductive argument that had existed since the time of Thales, two and a half millennia ago (Chapter 5). In the justice system, it's enough to prove 'beyond reasonable doubt' that someone is guilty or innocent, but mathematics requires a mathematical proof: starting from a simple premise, and building on that, step by step, until a conclusion is reached that shows *without* any doubt that something we conjecture about is either true or not. Only then could we say that the case is closed. And, unless there is some mistake, these proofs hold forever.

We've mentioned before about it sometimes having not been the right time for certain mathematical discoveries to be made, and this was the case with Fermat's last theorem. It needed other mathematics to be developed. It was like a puzzle with a few missing pieces. Now it was time for those pieces to be found, and an expert puzzler to slot everything together.

At Princeton, Wiles started working on the theory of *elliptic curves*. These were quite fashionable at the time, and extremely important, as they were showing something quite extraordinary. Elliptic curves have nothing to do with ellipses. The world of elliptic curves deals with forms such as $y^2 = x^3 + ax + b$, for some values of a and b (coefficients). For example, if we choose $a = 1$ and $b = 0$, we get $y^2 = x^3 - x$, and can plot it like so.

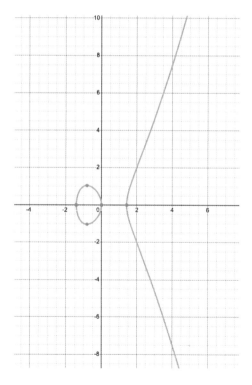

As usual, it is best to draw the curve so you can see what's going on beyond the equation. There are two shapes, one an almost perfect circle, and the other which is a curvy line. That curvy line goes to infinity in two directions along the y-axis: to $-\infty$ in one direction and $+\infty$ in the other. What happens at infinity? Obviously we don't *know*, but we tend to imagine (in mathematics) that, because at those points at infinity the curve goes in both directions, the points jump out of the second dimension and join up together. This line would then also become a circular shape. This is all topologically speaking – we can't reach infinity, but in mathematics and topology we can imagine it and make conclusions that help us to progress. And so, topologically, we can say that if we treat x and y as complex numbers, the lines that we get on the graph above represent two lines cutting a doughnut.

On the other side of the world, Japanese mathematicians had also been working on this type of mathematics. Goro Shimura (1930–2019) and his friend Yutaka Taniyama (1927–58) had started looking at elliptic curves and other similar things all the way back in the 1950s. Taniyama was a brilliant mathematician, but as Shimura once pointed out, he used to make a lot of mistakes. But these were, Shimura said, *good* mistakes, leading him in a good direction so eventually he arrived at the correct answers. Shimura thought this was an excellent tactic, and so he too tried to make good mistakes. But he discovered that it is actually very difficult to make productive mistakes in mathematics!

The two men worked on *modular forms*. Modular forms are functions that don't change (are invariant) under some actions with respect to groups. They also arise as differentials (the opposite of integration) in certain cases for elliptic curves. These are very complex functions, and they deal with complex numbers (in the upper-half plane of complex numbers, where the imaginary part is positive). The reason they are so interesting is that they are extraordinarily symmetric. There are so many internal symmetries in the modular forms that it sometimes seems like they can't possibly exist. But they do.

In 1955, at an international symposium at Tokyo, Taniyama posed two problems. They were in effect about elliptic curves and modular forms. Following this, Taniyama and Shimura worked together and conjectured like this: modular forms are actually elliptic curves in some kind of disguise. The conjecture, which became known as the *Taniyama–Shimura conjecture*, claimed that every rational elliptic curve is in fact a modular form.

Let's try to unpick this a bit. Their conjecture stated that elliptic curves over the field of rational numbers are related to modular forms in a certain way. We know what rational numbers are: numbers that can be expressed as a fraction of two whole numbers, and where the denominator cannot be zero. There are a number of properties which apply to this group of numbers: addition, multiplication, inverse element, associativity and commutativity (recall our description of groups in Chapter 24). Taniyama and Shimura were saying that there is a precise relationship between modular forms, with all their internal symmetries and periodic behaviour, and elliptic curves, that describe the solutions of a type of polynomial equations. All elliptic curves are, in a word, modular forms.

The Taniyama–Shimura conjecture was just that – an unproven conjecture – and originally it seemed a very, very far-fetched idea. But in time mathematicians around the world started believing that it was right, and they made new conjectures based on it. A significant amount of mathematics was now invested in the Taniyama–Shimura conjecture; if it was proven to be incorrect, all these other new assumptions would be wrong too. Because of this, the conjecture became even more important. Unfortunately, Taniyama died young and didn't live to see his jointly authored conjecture taking on a life of its own. For the time being, it was one of the great unproven conjectures of modern mathematics.

In 1985, the German mathematician Gerhard Frey (born 1944) made a crucial link between the Taniyama–Shimura conjecture and Fermat's Last Theorem. He proposed that if he could construct an elliptic curve from Fermat's equation (using an assumed 'Fermat triple' for the values of a, b and c, and a power greater than 2), then it would provide a solution to that theorem and prove Fermat wrong: that is, there *would* be a whole-number triple that would satisfy the equation raised to a power greater than 2. But that curve, because it was elliptic, would also, according to the Taniyama–Shimura conjecture, have to be modular, because they stated that *all* elliptic curves were modular. In other words, if there *was* a counterexample to Fermat's last theorem – provided by Frey's elliptic curve – then the Taniyama–Shimura conjecture would also be wrong. This in itself would eventually become known as the *epsilon conjecture*. Ken Ribet (born 1948), a mathematician at the University of California, Berkeley, came up with it in 1986 (and it hence became known after him, as Ribet's theorem). He showed that Frey's curve couldn't be constructed. This still didn't prove Fermat's last theorem, or the Taniyama–Shimura conjecture, but it showed that *they were right*, and that anyone who proved the latter would also bag the former.

Andrew Wiles learnt of Ribet's findings over tea one evening. He described it as an electrifying moment: suddenly he knew that in order to prove Fermat's last theorem, he just had to prove the Taniyama–Shimura conjecture. That 'just' is a bit misleading though – this was still an incredibly difficult thing to try to do! In fact, Ken Ribet once said that Wiles was 'probably one of the few people on Earth who had the audacity to dream that he can actually go and prove this conjecture'. Wiles abandoned all his other research and concentrated solely on this. It was in many ways a lonely project. No one else was working on it with such focus, if at all, so he couldn't discuss it with anyone. And there was such intense interest in Fermat's last theorem that, he said, the pressure of all the attention would have destroyed his concentration. He holed himself up and worked on it in secret. Seven years later, Wiles re-emerged from his study with the proof.

Wiles drew on findings from other mathematicians, and used Galois theory (Chapter 24) as an intermediary step. He reduced the problem by showing that elliptic curves are modular by counting them. The first step was to count modular forms, which took about three years! And then, finally, things started to fall into place. He gave some lectures in 1993 and announced at the end that he had a proof of Fermat's last theorem. He had proved the Taniyama–Shimura conjecture, not fully, but enough: he had found that all elliptic curves defined over the rational numbers are modular. And Fermat's last theorem followed from this. But when Wiles' proof was written up for publication, an error was found. Back to the drawing board went Wiles, but this time in the full glare of publicity, which he found very uncomfortable. After another year of working on it, and helped by his student Richard Taylor, Wiles was close to giving up when suddenly and unexpectedly he had what he said was an incredible revelation. It was a simple, elegant, beautiful proof. He couldn't believe he had done it. He looked at it, left his desk to walk around the department and come back to it to check it was still there. It was. His paper *Modular Elliptic Curves and Fermat's Last Theorem* was published in 1995.

Proving Fermat's last theorem was an achievement of incalculable importance in the field. The entire international mathematical community celebrated. Wiles would have been a shoo-in for the Fields Medal, but he was forty-one years of age when he proved the theorem in 1994, so didn't qualify. Instead, the International Mathematical Union awarded him a silver plaque in recognition of his achievements. It was just one of many honours and prizes to follow, and Wiles' basic strategy has already inspired decades of further work.

For Wiles, he recognised the immense privilege of having been able to work his whole adult life on his childhood dream. Almost no one in the world had thought it could be done. He was possibly the *only* mathematician who could have done it, with not only astounding technical brilliance in number theory but a huge and expansive knowledge of other mathematical fields, many of which he drew on in his proof. He was in the right place at the right time, certainly, able to use very new mathematics to solve this very old problem. But he was also a mathematician of incredible tenacity and persistence – a hero and an inspiration to anyone who wrestles with the fundamental problems of mathematics.

Even so, there are still mathematicians who believe that there was, after all, a proof provided by Fermat that he couldn't quite fit into his margin all those centuries ago, but that was much shorter than the proof we have now. Wouldn't it be amazing, if it did indeed exist, to know what it was, at last?

CHAPTER 37

Maryam's Magic Wand

In 1977, a girl was born in the Iranian capital, Tehran. She grew up amidst the Iran–Iraq war, and times were hard. She was clever, and dreamed of becoming a writer, inventing stories about girls who achieved great things and travelled the world. Little did she realise that she would soon be one of them, and mathematics would take her around the world on the adventure of a lifetime.

Maryam Mirzakhani (1977–2017) attended a middle and then a high school for gifted girls, one of many established by governments in the East and West for 'exceptionally talented' pupils. Her great talent began to show. It became almost a bit of a joke when she started giving more than one solution to the problems she was given to solve. One solution wasn't enough for her. She would try as many different solutions to a problem as possible, and then show them to her friends and teachers. Although it was obvious that she was naturally gifted, everyone who knew her at that time said that she also worked very hard, practised mathematics all the time, and just enjoyed doing it. Sometimes she would give herself a break and read books for a whole day – and then she would go back to mathematics.

One day at high school, she and her best friend Roya Beheshti (who also later became a mathematician) stumbled across a copy of six problems that had been set for the International Mathematical Olympiad. The idea for an international competition where young people would compete in mathematics had originated in the 1950s. At this time the world was very much divided into East and West, and the competition was originally for those countries under the Soviet bloc of influence, with the first International Mathematics Olympiad held in Romania in 1959. But other countries soon joined in. These Olympiads were quite exciting. Competitors would get to travel to whichever nation was hosting the competition that year, meeting kids from countries around the world, and being challenged by difficult problems they had never come across before.

Maryam found that she could solve three of those six problems she came across that day in school. She and Roya convinced their headteacher to put on special problem-solving mathematics classes for them, something that was commonplace at schools for talented boys, but not for girls. No girls had ever competed on the Iranian Mathematical Olympiad team either. But Maryam was instilled with a confidence that she could do it, even though she might be the first to do so, which she took with her through her life.

One of the teachers on the preparatory programme introduced Maryam to *combinatorics*, which is about mathematical structures and the laws that govern them, including counting and arranging (Chapter 31). Working through some interesting problems with her, it wasn't long before the teacher realised that Maryam brought something new to the whole area that he was teaching her about! *Graph decomposition* is a procedure in combinatorics by which a complex graph is broken down into mutually exclusive sub-graphs so that the original can be understood better. Maryam was still a high-school student when she jointly published, with her teacher, her first paper on this in 1994. But this was just the beginning of her extraordinary understanding of modern mathematics.

That same year, Maryam and Roya became the first girls to compete for the Iranian team at the International Mathematical Olympiad, held in Hong Kong. Maryam scored 41 out of 42 and won a gold medal. That wasn't good enough for her. She made the team next year and came away from the Olympiad in Toronto with a perfect score, the first Iranian to do so. After school she went to the Sharif University of Technology in Tehran, the best university in the country for STEM subjects, and while still an undergraduate attended some postgraduate mathematics courses too. She began training high-school students for participation at the IMO, and wrote a book to help them become better at problem solving. And her under-graduate work got her noticed by the American Mathematical Society. Harvard welcomed her to study for her doctorate in mathematics in 1999.

Now she became really interested in surfaces. When she was at Harvard University, she addressed a challenging problem that mathematicians had grappled with for years: determining the volumes of moduli spaces of curves (these are geometric entities whose points represent various hyperbolic surfaces). These surfaces can take unusual shapes, sometimes resembling doughnuts or amoebas. Mathematicians tried to calculate the volumes of all possible variations of these forms and failed. But Mirzakhani intro-duced a novel approach to this problem by drawing loops on the surfaces and calculating their lengths.

She turned to *Riemann surfaces*. A Riemann surface is a geomet-rical object that looks like, or can be described as, the basic plane of complex numbers. (We had a quick look at complex numbers and Riemann's work on manifolds in Chapters 26 and 29.) Riemann surfaces are multi-layered topological spaces, and have characteris-tics that we need to understand. One of them is the *genus*. This is a topological property that determines the Riemann surface's struc-ture; in its simplest terms, it's the number of 'holes' in an object. For example, the genus of a sphere is 0 and the genus of a torus is 1. (You may recall looking at Betti numbers in Chapter 32: they are related, but not the same as the genus.) How geometries differ for different surfaces depending on their genus is very interesting. A sphere with genus 0 has a positive curvature. The geometry on it is known as spherical, and, if you were to draw a triangle on the surface of the sphere, its angles would add up to more than 180°.

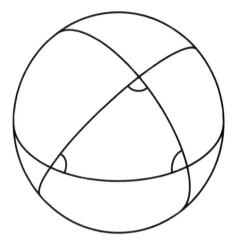

Now you're an old hand at this topology lark, this might sound familiar! Let's now look at the geometry of a genus greater than 0. A torus (a one-holed doughnut) has genus 1. In fact, a torus is considered to be a Euclidean plane, where rules of *flat* geometry are valid. How can that be? Because you can make a torus from a two-dimensional rectangle or square. Roll it to bring one pair of opposite parallel sides together. Now glue them, and you have a cylinder. Because we are working in topology and you can also squish and bend, the two open ends of the cylinder can be bent round until they are brought together. Glue them again, and you have a torus!

What Mirzakhani was interested in was the *geodesics* on such surfaces. Geodesics are curves that represent the shortest path between two points on a surface. Imagine that the Earth is a Riemann surface, and it is idealised as a geometrical object, a sphere. Of course, it is not a perfect sphere, there are so many imperfections on the grand and on the minute scale, but generally speaking we can say that Earth, as a mathematical model, is a sphere. If you told me that you were going to circumnavigate the Earth in a big circle, I would believe you could do it; you would have to change your mode of transport, but in theory you could go in one direction, make a full circle and return from the opposite direction to your starting point. This curve that you would walk is called a *closed geodesic*. But Mirzakhani was not working spheres, but on the geodesics on a torus.

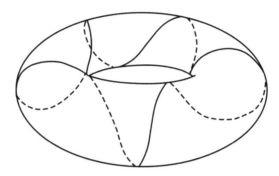

And from there, she looked at the geodesics on a *hyperbolic* torus. So instead of this smooth, equally distributed, easy torus, imagine if you can that your torus looks everywhere like a hyperbolic plane: the surfaces Mirzakhani was studying looked a bit like pairs of many-legged trousers. She also looked at surfaces which were like a punctured torus. In fact she was interested in most tori! She wanted to know how they could be classified.

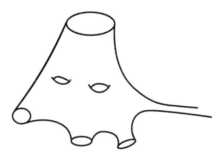

Mirzakhani started investigating what geodesics would look like on such surfaces. What happens on a surface such as a torus with more than one hole, or a hyperbolic torus, and walk from a starting point in a straight line? Your course would be different for different surfaces. As we saw, on a sphere that would be a simple circle. The path is closed, it closes on itself. If you did something similar on a torus, walked across, perpendicular to the central line of the torus and towards the centre, you would also describe a closed circle, and it would give you a closed geodesic. But imagine that you don't have much sense of direction, or you just want to start walking and go straight ahead without paying attention to the direction you're travelling in. This may end up being an infinitely long walk, and you would never get back to the place from which you started.

That's an interesting way of classifying surfaces, by the geodesics that are possible on them! There are then several things Mirzakhani looked at from there. One was about the curves that are described by such a walk. She looked at all the paths one can choose that would end up in a geodesic that was a *simple closed curve*. Such a curve does not cross itself and ends up in the place in which it started. You can categorise such curves. It's easy to do so on a sphere and an ordinary Euclidean torus. But on a hyperbolic torus or a torus with more than one hole it's a bit trickier, and that gave her some trouble. She discovered a way to simplify things.

Imagine that these are not at all curved surfaces for a moment. Imagine you are dealing just with a limited area of a flat plane, like a billiard table. (Mirzakhani used precisely this metaphor in her lectures.) These billiard tables didn't have to be rectangles – they could be of all different parallelograms. Imagine now that our billiard table is surrounded by mirrors. You hit the ball, and as it bounces off an edge, you could imagine the mirror showing where the ball would continue to go. Maryam therefore *extended* her billiard tables with these imaginary mirrors, and let the ball run, until the shape and the position of the ball path on the mirror table would be exactly the same as it was on the original table. The only difference would be that this imaginary billiard table would be translated from the original one, shifted along.

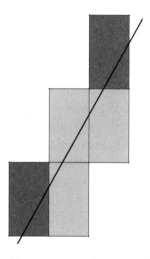

Now, the billiard table is a rectangle – a Euclidean plane – and recall that trick for constructing a torus from one of these that we saw earlier? If you take all the images of the billiard table, from the first one *until* the same one reappears translated, you can move them around to arrange into a rectangle-like shape in the Euclidean plane. There will be a pattern of original table, then some further billiard tables, until the first one is repeated. Move, stretch, squish, glue, and you get a Euclidean torus. Mirzakhani was able then to calculate very precise geometry for her surfaces – her billiard tables – and consequently they held for the tori.

Now she was able to study the characteristics of these surfaces of hyperbolic tori and determine their differences and similarities. She made an *imaginary map* where she outlined what all these surfaces were according to their characteristics. Two doughnuts with similar characteristics are close together on such a map, and if they are not that similar, they are far apart. Using this map, we could say what kind of torus whatever torus we're looking at is, and with what characteristics.

Mirzakhani and her friend Alex Eskin (born 1965) were studying such complex surfaces when they realised that there is some affinity between certain types of surfaces. They came up with something that is now called the *magic wand theorem*. It's in an area of mathematics not easy to explain, but let's try. Consider that the points in this special map we mentioned above each refer to some special surface. If you could wave a magic wand at a particular point on this map, it would bring about all the different surfaces that have an affinity with each other. You would be able to find out the characteristics of different surfaces with ease – as if, magically, all these details appear to you. Their theorem was this so-called wand that, when you waved it at a map of such surfaces, it would show you the characteristics of the surface at which you have pointed. Another metaphor that Eskin uses to describe a consequence of the theorem is a mirrored room. It doesn't have to be a square or rectangular room – almost any polygonal shape will do. Now put a lit candle in the middle of it, shining light in every direction. Will there be any spots in the room that are not illuminated? A side-effect of the magic wand theorem proves that there won't be.

Mirzakhani was a brilliant mathematician. In 2014 she became the first woman to be awarded the Fields Medal, and the first Iranian. But these were by no means her only 'firsts'. From her pioneering appearances and triumphs in the International Mathematics Olympiads as a schoolgirl to being the first Iranian woman elected to the National Academy of Sciences, she was first in many ways, and an inspiration to many mathematicians known and unknown to her. She died of cancer, aged just forty, but left behind a pathbreaking legacy – not just of outstanding mathematical work, but of the championing of girls and women to enter and excel in the world of mathematics.

CHAPTER 38

Patterns of Mathematical Success

There is a famous photograph of a ten-year-old boy sat next to the great Paul Erdős (Chapter 31) in 1985. They're both scratching their chins while looking over some mathematics. Erdős was indeed a friend to many, but the most prolific mathematician of his time was not just going out of his way to kindly explain something to a befuddled youngster. They were seriously discussing a mathematical problem.

That boy was Terence Tao (born 1975), today one of the most important living mathematicians. He was the youngest ever participant of that International Mathematical Olympiad, taking part in the 1986 tournament at the age of ten, and winning a bronze medal; the next year he won silver, and the year after, he won gold, at the age of thirteen. He's still the youngest winner of each of the three medals in the Olympiad's history. He completed his undergraduate studies in mathematics when he was sixteen, got his doctorate at twenty-one, and became the youngest full professor of mathematics (at the University of California, Los Angeles) by the age of twenty-four. Well, you might be thinking, I might as well just give up now. But we'll hear Tao's salutary advice about doing mathematics at the end of the chapter.

Tao has an extraordinary mathematical knowledge that is both broad and deep, and his field of contribution is very large. Like his childhood partner Erdős, Tao has many collaborators and mathematical friends from around the world, and has already written more than 300 papers co-authored with some of them. Let's look at some mathematical discoveries he has made so far.

In 2004 he and his friend Ben Green (born 1977) solved a problem related to the *twin prime conjecture*. Twin primes are prime numbers that are 'next to' each other, the difference between them being less than or equal to 2. You can easily find examples among smaller numbers (5 and 7, 17 and 19), but as the numbers get larger, the primes become increasingly rare. The twin prime conjecture says that there may not be infinitely many twin primes. Now, this is quite extraordinary: there are infinitely many primes, but how would we show whether there are or are not infinitely many twin primes? This remains an unsolved problem.

Instead, Tao and Green started looking at slightly different patterns in the prime numbers – prime number progressions, where the primes are equally spaced. For example, the primes 3, 7 and 11 have a spacing of 4. They're not twin primes, but they're close. The next prime is 17, so the progression doesn't hold any further. It's just a little three-term arithmetic sequence of prime numbers with a common difference. Now, the *Green–Tao theorem* states that it is *always* possible to find a progression of prime numbers in an arithmetic sequence, of equal spacing and of some length. You may have to search the infinity of integers for a long while, but you will find it. And considering how little we know about patterns in prime numbers, this is an important theorem.

Another theorem that Tao worked on was actually conjectured by Erdős around 1932. It relates to the *sign sequence*, or *bipolar sequence*. This is a sequence of alternating positive and negative 1s (1, -1, 1, -1, 1, . . .) that go on indefinitely. The sum of this sequence would be, you'd think, 0: the 1s would cancel the -1s into infinity.

The problem Erdős posed was about what happens if you started skipping some numbers in this sequence, and stopped at some point. Erdős conjectured that for *some* number of jumps (say, 100 jumps) and for *some* length of these jumps (so you jump one, two, three or more numbers at a time), you will get a sum that is different from 0. In fact, he conjectured that it will be larger than some number C. This is called the *Erdős discrepancy problem* because at its centre is the difference between C and 0. That is the discrepancy – the value of how far C is away from 0.

If you think about it, depending on the length of jumps, you could end up with only positive numbers in your sequence, or only negative. And depending on the number of jumps you take, that number C could be not very large at all, or it could be huge. It all depends on what jumps you choose, and the length of your sequence. The Erdős discrepancy problem asked for proof or disproof of this conjecture – that it's possible to find positive values for the intervals (the length of the jumps) and the number of terms in your sequence so that the sum of the sequence would be greater than C.

Eighty years after Erdős posed this problem, Tao proved that for some sequence and for different lengths and numbers of jumps, there will always be *some* sum which is far away from 0. The master problem-deviser Erdős had established monetary prizes for solving his puzzles in his lifetime, and this is still being administered after his death (search for 'Erdős Problems' if you'd like to try your hand at some of the hundreds remaining). This particular problem had a $500 award attached to it. But Tao won't be cashing the prize. Very few people would actually do that, he said; a framed Erdős cheque above your desk is worth far more.

There was also a very famous, and very old, problem that Tao tackled and solved with a mathematician friend, Rachel Greenfeld. It was about tiling – how particular shapes can cover an entire plane without gaps or overlaps. This was something that mathematicians and geometers had wondered about since antiquity, and artists and architects had created pleasing patterns for walls and floors using such tilings. This kind of tiling is *periodic*, where the tiles cover a plane in a uniform way. We can use translation, rotation or reflection, and we can also use one or more different tiles to cover the plane. And were you to look from above at any part of this plane expanding indefinitely in all directions, you would see the exact same pattern as anywhere else.

The ways in which we can tile a plane have been investigated since Kepler (and we'll see more on his thinking about honeycombs in the next chapter). Fast forward two hundred years, and a Russian chemist showed in 1891 that *every* periodic tiling of a plane can feature one of seventeen groups of *isometries*. Isometry is another word for mathematically describing symmetry. It's when we preserve the distances between shapes when we move them around in some way. Mathematically, that means isometry is a *distance-preserving map* – we move a tile and it is congruent (exactly the same as it was) to itself, just in a different place. And why was it a chemist who discovered that all the possible tiling patterns that people have made anywhere, and at any time, fall into one of seventeen groups of isometries? Because chemical molecules, it turns out, also have symmetrical structures, and knowing the ways they can map onto each other is as important to chemistry as the geometry of tiling is to mathematics.

This was all well and good until the 1960s when mathematicians discovered weird sets of tiles that covered the plane, but *aperiodically*. As the name suggests, you can still tile the whole plane, going into infinity in all directions, but only in ways that never repeat. The tiling is not periodic; you can't translate a section from one place to another. Every section of the plane you look at has a slightly different arrangement, although there is order in it. The challenge now was to try to understand the structure of these aperiodic tilings.

One of the first aperiodic patterns to be created used a set of more than 20,000 tiles in order to avoid repetition. Such a huge number wasn't very mathematically satisfying, and in the 1970s, the British mathematician-physicist Roger Penrose (born 1931) improved upon it, discovering in 1974 that with just two tiles, 'kites' and 'darts', one can aperiodically cover a plane. (We'll come back to Penrose in Chapter 40.)

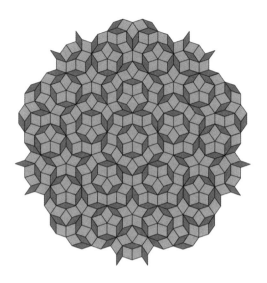

Penrose tiling has actually made appearances in the real world. You can see it on the floor at the entrance to the Mathematical Institute in Oxford and in the foyer of the Mitchell Institute at Texas A&M University, for instance. But when the pattern was used by a manufacturer of toilet paper to emboss the sheets in a pleasing pattern in the 1990s, Penrose promptly took legal action. He had patented the design, they hadn't sought his permission, and he was presumably a bit grumpy about what they envisaged his beautiful mathematical discovery should be used for!

Could Penrose tiling be improved upon? Since aperiodic two-tiling had been discovered, the question was now whether there was a *single* tile that could do the same thing. The race was on to find that aperiodic monotile, which was playfully called an 'einstein' – not after the great man himself, although mathematicians would devote an Einsteinian amount of thinking to it, but after the German *ein Stein*, or 'one stone'. The answer to the question of whether an aperiodic monotile existed was, at that time, both yes and no. Yes, if you're allowed to rotate and reflect the tile, and if it's allowed to leave gaps, at least for a while – they eventually get covered by other rotated or reflected copies of the tile until the whole plane is covered. But the answer is no if you're *not* allowed to rotate the tile, but only use translations – it was proved impossible to cover a plane aperiodically in that way. Over the years, mathematicians discovered shapes could aperiodically cover the plane if they overlapped, or found shapes with disconnected pieces that would do the job. But it seemed there was no simple, truly two-dimensional tile that would fit the bill.

Mathematicians, being mathematicians, became interested in finding out if this was the case in higher dimensions too. If there is no singular aperiodic two-dimensional tile that fills an infinite plane, they supposed that there would also be no three-dimensional block, or higher-dimensional more complicated shape (whatever that would be), that would completely fill the dimensional space aperiodically. This became known as the *periodic tiling conjecture*.

Here's where Terence Tao and Rachel Greenfeld came in. In a surprise to the mathematical community, and perhaps even to themselves, they *disproved* this conjecture in 2022. They managed to construct a tile that *could* fill a very high-dimensional space aperiodically, but not periodically, using translations (no rotation or reflection). Tao and Greenfeld had to come up with a new mathematical language to reframe the problem and worked through some very complex mathematical procedures to get there. It became a question of logically encoding certain constraints through their equations which they could then populate their grid – they likened it to a game of sudoku. Trying to imagine what their resultant tile would actually look like and how it works in the size of the dimensions they were talking about frankly boggles the mind. Tao described it as 'a nasty tile' – it's full of complicated twists and holes. The dimension it exists in is so large that were you to try to write the number down, even if you used all the books in the whole world, you'd run out of pages. But, there was the proof: it exists.

Very soon after that, in 2023, the mathematics community was stunned again. A true einstein was discovered not by a professional mathematician, but by an English mathematics enthusiast and retired print technician called David Smith. He was playing around with an odd thirteen-sided shape that he had constructed using a computer program and found that it tiled his whole screen with a pattern that didn't seem to repeat itself. He moved to his table, cutting dozens of copies of the shape out of card and assembling them together. They were tessellating in a way he hadn't seen before. He told a computer scientist friend about it and, with two further researchers, they investigated it. Within a couple of months they had the proof that their tile, simply dubbed 'the hat', because that is what it looks like, was the aperiodic monotile that mathematicians had spent the last fifty years searching for.

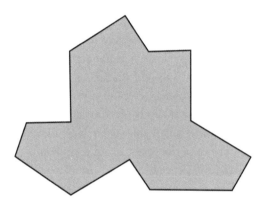

And in fact, Smith had already found another one! This one he named 'the turtle', and it helped the team realise that both shapes actually belong to an infinite family of aperiodic tiles that are all einsteins. In the process, they had also discovered a brand new way of proving aperiodicity.

This was undoubtedly a 'eureka' moment – a kind of effortlessly brilliant revelation that redefines a field. Perhaps we've imagined all mathematical discoveries to be something like this. But it's not so. Terence Tao, widely hailed as a genius from such a young age, is careful to point this out. And this might make us feel a bit better about our own chances when engaging with mathematics.

Tao once said that when a problem is solved, all people remember is the solution. To the outside world, then, mathematics can appear to be a straightforward process of talented people making the right choices and following the correct sequence of steps to reach their goal. But in reality, many mathematicians wander through the landscape of mathematics blundering, wasting time following dead ends, and having embarrassingly bad ideas. Of course, history chooses not to record that, and nor do many mathematicians! For Tao, any 'eureka' moment is more like a 'hitting your head' moment, when you kick yourself for not noticing something before. And if a great mathematician like Tao can feel that sometimes, then we shouldn't feel discouraged if we don't get it immediately either.

CHAPTER 39

Packing for a Trip to New Dimensions

Are you one of those people who always wants to take more things on holiday than they can fit in their suitcase? After a few failed attempts to close the zips, you'll have a pretty good idea of the maximum amount of stuff you can fit in. But it wouldn't be a mathematically precise idea, for despite your very best sweaty efforts, there will always be some empty space remaining in your luggage that you won't be able to fill, however hard you try.

The problem of packing has been a concern of mathematicians for many centuries too. We're not talking about luggage here, but rather Euclidean space, and how it can be packed in the most effective, optimal way. Kepler first thought about it in 1611 when, one late winter evening, he was walking in Prague as snowflakes were falling on this coat. He looked at the patterns they made, and realised that these delicate shapes were all of the same overall shape. They differed in intricate little irregularities, but they were all hexagonal. It got him thinking about the snowflakes and their coverage on a plane, and from there he started investigating *packing* in a plane in two and three dimensions. Space was then, mathematically speaking, considered only Euclidean. We now know that it is possible to imagine and prove the existence of other types of spaces in mathematics. Nevertheless, when we talk about packing, we still consider the Euclidean model. But surely there's not much in packing that mathematics can reveal? There is, and in fact this has been one of the most famous recent discoveries in mathematics. Let's look at how it works first.

In two dimensions the packing problem is about packing circles. In three dimensions, it's about sphere-packing. The aim of doing this would be to pack as many spheres of the same size as possible in some amount of space. Here comes that problem of *waste* – the empty space between the spheres. You won't be able to pack the entire three-dimensional space with spheres, but you're going to do your best to achieve the densest packing possible. You can work out how well or badly you're doing by calculating the volume of the spheres and the volume of the space in which they are packed to work out the ratio of these two volumes: this is the *packing density*. The higher the packing density, the more space is being filled.

It's said that Sir Walter Raleigh, Queen Elizabeth I's explorer for new colonies, wanted to find out what was the best way to pack cannonballs on his ships. He wanted, of course, to carry the maximum amount in the minimum possible space. He asked Thomas Harriot (1560–1621), a mathematician and astronomer who worked for him at the time, to solve this problem. Harriot in turn wrote to Kepler to ask for his help. In 1611 Kepler said that no arrangement of equally sized spheres is better than that of hexagonally packed spheres. Say you ignored Kepler and packed your spheres any which way; you'd achieve something like a packing density of $\frac{2}{3}$ in three dimensions. But if you followed his suggestion and packed spheres in a hexagonal formation, then the density would be about 10% greater. This was the *Kepler conjecture* (sometimes the *honeycomb conjecture*). It seems obvious enough, but it remained unproved for centuries.

Over the years, various mathematicians made attempts at arriving at a proof. Friedrich Gauss (Chapter 23) showed that

Kepler's arrangements were the best possible gridlike arrangements, but couldn't prove that another, more irregular arrangement wouldn't possibly do better. But then in 1998, Thomas Hales (born 1958) gave a preliminary proof of the Kepler conjecture. This was a 'proof by exhaustion' – done by creating a code or program that checks for all the possible cases to show that your hypothesis is right. Computers were by this point an increasingly integral part of doing (some kinds of) mathematics, and it was becoming common practice to make use of computers to prove theorems if they could prove helpful in crunching through tedious complex calculations. Hales' long proof was submitted to journal referees who, four years later, and despite not having been able to verify all of the computed calculations, accepted the proof, being '99% certain' it was valid. That didn't stop Hales. He marshalled more computer power to provide a formal proof that could be verified by automated proof checking software, and that final formal proof was accepted in 2017.

Kepler was right: the maximum packing density is achieved when you pack in a hexagonal way. In two dimensions, that would mean six circles are placed with their centres in the vertices of a hexagon, and a seventh placed inside, with all their sides touching. In three dimensions, you'd follow the same pattern for your bottom layer of spheres, with the next layer placed in such a way that the spheres lie on top of the hollow spaces between the first layer's spheres, making a pyramidal shape. And so on. Although that seems obvious – we've all seen fruit piled up like this at the greengrocers – there are actually different choices you can make about arranging spheres on subsequent layers. The best packing density you can achieve is $\frac{\pi}{3\sqrt{2}}$, which comes close to 0.74 with a never-ending, non-repeatable decimal part.

A surprising application of the mathematics of sphere-packing comes in communication theory, particularly related to radio communication. The mathematics behind sphere-packing can be applied to codes that correct for errors from other signal interferences, or to encode information before it is transmitted. In order for a communication to be clear, there needs to be this close packing – no space for anything in between, and no overlapping. If you have a signal represented by a sphere 1, and a signal with a sphere 2, clear communication is when they are close together but do not overlap – the latter case leads to the potential for confusion.

So the Kepler conjecture was proved. But the problem wasn't quite over yet. Now, twenty-first-century mathematicians were interested in higher dimensions.

As we've seen in earlier chapters, it's difficult to imagine higher dimensions. But why all the fascination with them? Well, working with higher dimensions in mathematics can provide us with different models for things that happen in reality. With the three dimensions of Euclidean space, yes, we can model space and how it looks. But if we add a fourth dimension and say it is time, we can see how that space changes over a duration. In a similar way, further dimensions can be made to represent different things that we may be interested in. There is this freedom with dimensions that can help us model different systems and their structures, and what can possibly happen in them or how likely it is that something we are interested in would happen. The three variables of three-dimensional space – x, y and z, each corresponding to an axis of the three-dimensional coordinate system – can be extended into much more complicated equations in much higher dimensions. The possibilities are literally endless. Depending on what you want to measure, or if you wish to see how different areas of a system interact, you can assign different meanings to additional dimensions. In fact, in mathematics, anything you can measure in n dimensions is a point in an n-dimensional space. In that way, you could see what is mathematically possible, and that can be a very useful model for real-life systems.

Although Hales' final proof wasn't ready until 2017, everyone was pretty much certain that what he had showed in 1998 was true. The honeycomb structure, which is a special case of a hexagonal *lattice*, was the best way to pack spheres. A lattice in mathematics can refer to a number of things, but in this context, it is about the arrangement of objects that are evenly spaced, with each point equally distant from all others.

Could a lattice of some kind hold for higher dimensions too, and help to discover the best way to pack spheres in the densest way possible in dimensions higher than three? It was pretty obvious that the findings for three-dimensional optimal sphere-packing were not going to lead directly to results in higher dimensions. In fact it's almost impossible to see what's going on when the number of dimensions is very large.

Mathematicians already knew of one formula that they could use as a starting point, and which holds for *any* dimension. If you can imagine a huge box into which you'll successively chuck your spheres wherever they'll fit, in whatever dimension you're working in, you will get *at least* $\frac{1}{2^d}$ of the box filled with spheres, where d is the number of dimensions of your space. This is the minimum packing density you can achieve. It sets a low bar – the very worst packer would still get this, however bad their packing system is. According to this formula, in three dimensions the minimum packing density is $\frac{1}{8}$, or 12.5% – much lower than the 74% that Kepler conjectured and Hales proved was possible. But this is an important formula nevertheless, as it's a rule that applies across all dimensions.

In the 2010s a remarkable mathematician set her sights on the problem of optimal sphere-packing in higher dimensions. Maryna Sergiivna Viazovska (born 1984) was born in Kyiv and completed her doctorate at the University of Bonn in 2013, working on modular forms. We met with modular forms earlier (Chapters 36 and 37): they're functions with very special symmetries which, if we illustrated them, might look something like M.C. Escher's famous circular tiling of angels and devils. Their symmetry makes them incredibly useful, as they are very effective in describing a whole host of mathematical laws.

Although she was still young and relatively inexperienced, Viazovska now brought her knowledge of modular forms to bear on the sphere-packing problem in dimensions of four and higher. All kinds of issues, obstructions and redundancies arise with the higher dimensions, but a few dimensions are easier to work with. These are the eighth and twenty-fourth dimensions. In collaboration with two other Ukrainian mathematicians, she zeroed in on these particular dimensions, which each have a highly symmetric arrangement, two lattices respectively called the E_8 and the Leech lattice. Into these lattices, spheres could be packed more densely than in other arrangements in other dimensions (such that mathematicians could find). Viazovska and her colleagues looked to modular forms to help them find a function which would allow them to match that kind of lattice. If they were successful, they would have, in a sense, a 'magic' function that would provide the densest packing. But only if the function could be constructed in the first place!

They made little progress, and the two men eventually moved on to other things. But Viazovska couldn't let this go. She felt the problem was made for her, with her particular interests and skillsets. Years passed, until 2016, when she published a paper revealing the new function for the eighth dimension. The optimal density for sphere-packing here is $\frac{\pi^4}{384}$. Very quickly following publication of this result, she was convinced to dive into solving the problem for the twenty-fourth dimension, which she and a few colleagues came up with after a very intensive week of work: it's $\frac{\pi^{12}}{12!}$. These are very simple formulae, but Viazovska had got to them through some extremely complicated mathematics. Viazovska was producing a type of work never seen before, yet she was still only a post-doctoral researcher in Berlin at the time. And her breakthrough paper in 2016 was her first on the subject. For someone so young and so unknown to the wider mathematical community at this time, it was an even more remarkable achievement.

So, we now know that optimal sphere-packing *can* and *does* take place in one, two, three, eight and twenty-four dimensions. We haven't got any general results for other high dimensions larger than three. Neither do mathematicians really know *why* this is happening yet, but it is. Viazovska also showed that certain lattices are optimal, or of greatest efficiency. The E_8 and Leech lattices were, mathematicians thought, not only the best ways of packing spheres but are the best arrangements for a range of applications, one of which is computer security and cryptography. Viazovska made a bold conjecture and then proved that these two lattices are universally optimal – their way of packing is universally efficient. Again, because we now know this, it can be applied in a range of contexts, such as the signal processing we looked at earlier in the chapter. This was not only an advancement in mathematics, but a great step forward for the sciences.

In the wake of her first paper on sphere-packing in 2016, Viazovska was herself surprised by how excited others were about her result. She thought they'd like it, but didn't dream she would receive so much attention for that work. It was enough to win her the Fields Medal in 2022, only the second woman to ever achieve this, after Maryam Mirzakhani. She completed a work that has lasted centuries. From Raleigh and Harriot, via Kepler and Hales, Viazovska has shown how to pack in higher dimensions too, allowing us to take a trip into multidimensional mathematical landscapes.

CHAPTER 40

Dreams of New Mathematics

There have been so many topics we've covered in this little history of mathematics. We've gone from tables and lists, fractions and proofs – as if laying the foundations of a building to come – to then see how, via continuously accruing an ever more modern language of mathematics, mathematicians developed methods and techniques to develop more complex ideas, building storey upon storey until now the structure pierces the skies above us and takes us into a world of other dimensions.

Mathematics is a science of patterns, a method of finding succinct, general, abstract laws that underline and describe all other structures, and which make this a discipline of its own. Mathematics is also an art, drawing on its authors' creativity and leaps of imagination, made for the pure love of the subject. Sometimes we immediately see applications of mathematical ideas, but sometimes not, at least not immediately. Mathematical ideas don't always come together with their 'useful' function. Mathematics is also expressive of a profound, precise beauty, and asking any research mathematician why they have started working on it will confirm this to you.

The story of mathematics gives us a unique history of the world. From prehistoric cave-dwellers inscribing a baboon bone to packing spheres in the twenty-fourth dimension, there's always been a need to understand and record patterns in whatever form of mathematical language you have at your disposal. And these messages of distinct knowledge have then been passed on to others. The language they've been written in has advanced enormously over the centuries, but the principles and basic ideas that underline it are quite similar. Recognise a pattern, record it in the most succinct, accurate and abstract way, and leave it for others to ponder on.

We have traced the long course of how mathematicians clarified things for themselves, and made sure that whatever they understood, they recorded in such a rigorous way that it could be used by anyone, perhaps born centuries later, or anywhere on the globe. That is the beauty of mathematical statements, so clear, rigorous and abstract that they can be deciphered and taken up by people across space and time. As we've seen, this has happened time and again, and will continue to do so.

Mathematical activity has taken many forms. The perhaps overly familiar idea of lone mathematicians holing themselves up to concentrate on abstract problems in isolation is only one tiny part of the story. Even lone mathematicians are in a way working with others, through looking at the old problems and solving them, hearing about an unsolved problem from their colleagues, or even from books on the history of mathematics. We've seen even very early mathematicians responding to the work of others, and doing and developing mathematics directly linked to their particular time and place in the world. From fairly local roots at its beginnings, mathematical knowledge was quickly translated and transmitted across borders. Its universal language and our globalised and interconnected world mean that today, the making of mathematical discoveries through international cooperation and communication is pretty much the norm. There are fractures in the world that may be raging, but mathematicians always seem to find ways to cooperate.

The extraordinary advances in technology have totally transformed the way mathematics has been done, too. We have seen how computers enabled mathematical discoveries to be made, be that through crunching enormous amounts of numerical data, speeding up calculations or checking proofs. Algorithms can develop our abilities. There's an increasing number of instances where we can see cutting-edge mathematics making use of the computer, something that's been around for less than a century. In turn, we have mathematics to thank for that technology in the first place. Today's burgeoning field of Artificial Intelligence (AI) may seem like magic, but behind it all, it's mathematics. The spotting of patterns, making decisions, abstracting data, creating networks – these are all processes and algorithms built into AI through mathematics.

The new horizons opened up by AI are being explored by mathematicians as much as anyone. There are funds and prizes that encourage mathematicians and computer scientists to develop new models that would allow AI to prove more things in mathematics. This is exciting, but some mathematicians are sceptical. One fear is, once AI can project conjectures and prove mathematical theorems, what will mathematicians do? Will the profession simply become boring, with all the heavy lifting done by computers? Will there be any need for mathematicians at all?

This strikes at the heart of big questions of creativity and understanding. You can already make a computer do things you haven't thought of before, and we can imagine that this will only increase as machine learning improves over the coming decades. Is this creativity? There is a tension here. AI can be programmed with a clear aim in mind – success, winning or completion is obvious when it's reached. But mathematics so often involves hunches and feelings, or follows byways or makes surprising links. Often the most beautiful *and* frustrating things about mathematics are where the creativity lies.

Mathematician Stephen Smale (born 1930) once said that the understanding of mathematics doesn't really come when you only hear it or read it. It is a start, and hopefully a good one, but it is not how you fully understand things. That comes, he said, from *rethinking* what you've heard or read. You have to digest the mathematics you've learnt by doing it, putting it into your own context, and only then do you understand it. (Smale also said that the best pieces of new mathematics he developed were those he thought of while he was on the beaches of Rio de Janeiro. He had a little problem with people giving him grants after that.) The Nobel Prize-winning physicist and mathematician Roger Penrose also talks about understanding in relation to consciousness – the awareness that makes us human, a universal quality, which is essential to understanding. His linking of quantum mechanics (which describes how small objects behave both as particles and waves) and consciousness is controversial. But he's clear that, although consciousness is closely connected to mathematics, it is *not* computational. The *understanding* of mathematics is very different to being able to do computations. The human understanding of any system of rules transcends that system. AI, at the moment at least, remains within it.

Just as Gödel and his incompleteness theorem(s) showed, no matter how complete and perfect the system appears, it will never be able to prove some things. We are able to understand mathematics, and keep doing it, *despite*, not *because of*, the supposed completeness of that mathematical system. Over and over again, we have seen the great advances in mathematics coming from a place of humans understanding what was going on, and creating new rules and new mathematics. And then of course, we mustn't forget that mathematicians are primarily problem-solvers. If AI itself becomes a problem, they will get on with solving it, there is no doubt about that.

What new mathematics might be on the cards in the near future? The topics left to be worked on and discovered are many. As we've seen, some of Hilbert's problems remain unanswered; Smale also came up with his list of problems for the twenty-first century. A quick internet search brings up a whole host of burning questions. Modular forms are interesting and difficult, and there are certainly many PhDs waiting to be done on these. We saw how the work in high dimensions is something that has taken off in a grand way since the nineteenth century, and there are unique and interesting new insights to be gleaned from seeing if theorems proven in the third dimension are valid in higher dimensions too. There are also fascinating and important applications of mathematical research in higher dimensions. In the social and political sciences, the possible interactions between people can be studied by mathematically modelling things such as their loyalties and beliefs. Game theory in higher dimensions may yet give us some new interesting solutions to achieving better equilibria than what we have managed hitherto. Such applications are of increasing importance in our globalised world.

History teaches us that as long as humans have existed, they have done some kind of mathematics – why would the future be different? It's not possible to stop doing mathematics. So what is it like to be a mathematician? There is a sense of joy and achievement that mathematicians seek, like that when you complete a puzzle. A mathematician needs persistence and an ability to get over failures without regret, and a certain amount of audacity in believing that, despite what everyone else thinks, you can solve a problem, no matter how long people might have been puzzled by it. Sometimes the doing of mathematics is rather joyous and funny, sometimes tantalising or even torturous. Just like musicians and sportspeople, mathematicians learn to enjoy both these extreme states and everything in between.

Remember how Kepler imagined a dream in which he oversaw all the things that were happening in the solar system by flying to the Moon? It gave him a different viewpoint from that he would have had if he'd stayed on Earth. From such a vantage point you could reorient yourself in order to see the true relationships in the heavens above and all their beautiful laws and motions. It took a translation in perspective to see this. Discovering some similarly new vistas provided by AI, we might be able to make new discoveries in mathematics *and* look anew at what the subject itself could be in the future. Perhaps new mathematicians will come up with similar 'dreams' to see afresh the vast landscapes of mathematics, and further, with the help of AI, be able to imagine how they can extend the currently known mathematical universe. Who knows. We can but dream of the possibilities.

Index

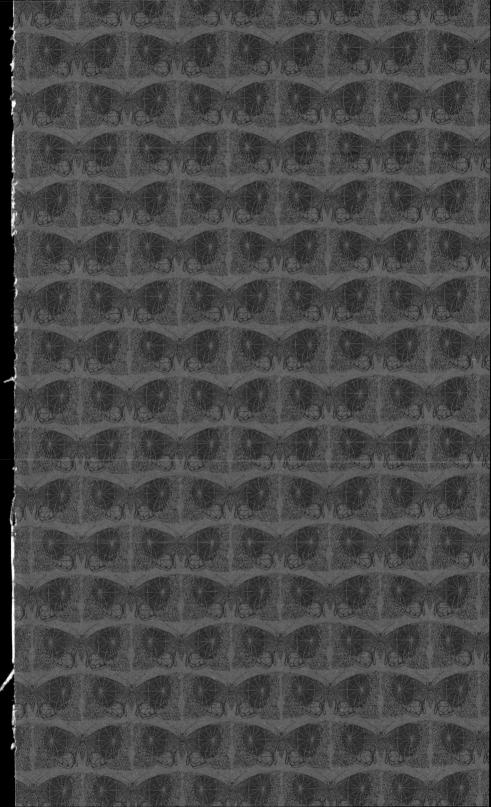